Blue and Green Water Resources and Their Sustainable Use: A Case Study of the Heihe River Basin, China

刘俊国　臧传富　曾　昭　著

马　强　徐广峰　郭凤春　苏立志　译

黄河水利出版社

·郑州·

Abstract

The Heihe River Basin is a typical arid region inland river basin. This book starts from the spatial and temporal dynamic distribution of blue-green water and the sustainability of water resources utilization in Heihe River Basin. The spatial and temporal dynamic distribution pattern of blue-green water in Heihe River Basin in recent 30 years, and introduced the concept of water footprint, the sustainability of the use of blue-green water in the basin was evaluated. Chapter 1 introduces the water resources evaluation in Heihe River Basin, and introduces the concept of water footprint. Chapter 2 introduces the research methods of this book. Chapter 3 introduces the spatial and temporal distribution pattern of blue-green water resources in Heihe River Basin. Chapter 4 introduces the impact of human activities. Chapter 5 introduces the temporal and spatial differences in typical years. Chapter 6 introduces the history of the evolution of the black and blue water in Heihe. Chapter 7 introduces the evaluation of water shortage in Heihe River. Chapter 8 presents the conclusions and prospects.

This book can be read for water resources management and planning and other relevant professionals, especially for the Heihe River Basin and China's arid and semi-arid watershed water resources management and socio-economic development planning for the research content of the staff to provide theoretical reference.

图书在版编目(CIP)数据

黑河流域蓝绿水资源及其可持续利用 = Blue and Green Water Resources and Their Sustainable Use: A Case Study of the Heihe River Basin, China:英文/刘俊国,臧传富,曾昭著;马强等译.—郑州:黄河水利出版社,2017.8

ISBN 978-7-5509-1822-1

Ⅰ.①黑… Ⅱ.①刘…②臧…③曾…④马… Ⅲ.①黑河-流域-水资源利用-研究-英文 Ⅳ.①TV213.9

中国版本图书馆CIP数据核字(2017)第208172号

组稿编辑:李洪良 电话:0371-66026352 E-mail:hongliang0013@163.com

出 版 社:黄河水利出版社　　网址:www.yrcp.com

地址:河南省郑州市顺河路黄委会综合楼14层　　邮政编码:450003

发行单位:黄河水利出版社

发行部电话:0371-66026940、66020550、66028024、66022620(传真)

E-mail:hhslcbs@126.com

承印单位:虎彩印艺股份有限公司

开本:787 mm×1 092 mm 1/16

印张:8.25

字数:230千字　　印数:1—1 000

版次:2017年8月第1版　　印次:2017年8月第1次印刷

定价:30.00元

Preface

Ensuring adequate supply of fresh water resources is essential not only for humans but also for ecosystems. Water can be divided into blue water and green water. Blue water mainly refers to rivers, rivers, lakes and shallow groundwater; green water is derived from precipitation, stored in unsaturated soil and the plant in the form of evapotranspiration absorption of the part of the water. In the past, people tend to pay more attention to blue water and ignore the green water. But green water plays a very important role in ensuring food production and maintaining ecosystem balance. Green water supports rain-fed agriculture. At the same time, it is also an important source of water for the coordination and balance of terrestrial ecosystems and plays an important role in maintaining ecosystem health.

In order to promote this concept to the world, we have translated the three professors in particular, in which Ma Qiang translated the first chapter of the introduction, the eight chapter of conclusion and prospect, and reference, and completed the manuscript of the book. Xu guangfeng translated the second chapter of overview of research areas and research methods, the five chapter of study on spatial and temporal difference of blue-green water in heihe river basin in typical years. Guo fengchun translated the three chapter of study on spatial and temporal dynamic distribution pattern of blue green water in the Heihe River Basin under natural conditions, the four chapter of the spatial and temporal dynamic distribution pattern of blue-green water in the Heihe River Basin under the influence of human activities. Su lizhi translated the six chapter of analysis on the evolution Trend of blue-green water in the Heihe River Basin, the seven chapter of evaluation and sustainable analysis of water shortage in the Heihe River Basin.

Foreword

Ensuring adequate supply of fresh water resources is essential not only for humans but also for ecosystems. Water can be divided into blue water and green water. Blue water mainly refers to rivers, lakes and shallow groundwater; green water is derived from precipitation, stored in unsaturated soil and the plant in the form of evapotranspiration absorption of the part of the water. In the past, people tend to pay more attention to blue water and ignore the green water. But green water plays a very important role in ensuring food production and maintaining ecosystem balance. Green water supports rain-fed agriculture. At the same time, it is also an important source of water for the coordination and balance of terrestrial ecosystems and plays an important role in maintaining ecosystem health. In recent years, blue-green water research has led to rethinking on the concept and evaluation of water resources in the scientific community, gradual ly affect the mode of thinking of human's water resources management, and become a hot topic in the field of hydrology and water resources research.

At the same time, with the rapid socio-economic development, water shortages has become the bottleneck of sustainable development in many countries and regions. Under the dual effects of climate change and human activities, the scope of the world and the extent of its impact of water shortage problem have increased further. Water shortage has become one of the major obstacles of global economic and social sustainable development. Study on spatial and temporal distribution characteristics of blue-green water resources, clarifies the evolution law and driving mechanism of blue-green water resources in changing environment, reveals the sustainability of water resources utilization, accurately identifies the shortage of water resources, causes and coping strategies, and has been affected by government departments, public and scientific research personnel of the widespread concern.

This book puts forward the theoretical framework and method of quantitative evaluation of blue-green water resources, and puts forward a comprehensive evaluation of water resources combined with water quantity and water quality, which is based on the major scientific problems of blue-green water resources assessment and the important practical needs of water resources shortage in changing environment. The book analyzes the spatial distribution pattern of blue-green water in the watershed, and reveals the evolution law and driving mechanism of blue-green water resources in the changing environment. Based on the water footprint concept and the water quality of the river basin, the water quality of the basin is evaluated and water quality of water shortages, and analysis of river basin water use sustainability. A typical study on the evaluation of blue-green water and water shortage in the study area was conducted. This study has developed the theory and method of blue-green water resources evaluation in changing environment, and provided scientific basis for the comprehensive response of water shortage in Heihe River Basin.

This book's academic thinking and writing framework is under the auspices of Professor Liu Junguo, Cheng Guodong academician gave a lot of guidance, and put forward valuable advice. This book is Professor Liu Junguo scientific research team collective efforts of the crystallization, but also its guidance of the master's hard work over the years the results of hard work. Among them, the chapter on the evaluation of blue-green water in Heihe River Basin is mainly completed by Dr. Zang Chuanfu; the Heihe River Basin water shortage and sustainable evaluation Chapter is mainly completed by Zeng Zhao. The preparation of the book is completed by Professor Liu Junguo; text editing and publishing matters is completed by Liu Junguo and Zang Chuanfu togetherly.

The project has been supported by the National Natural Science Foundation of China, a comprehensive research project integration project" Integrated Simulation and Prediction of Water-Eco-Economic System in Heihe River Basin" (project number: 91425303), integrated project "Heihe River Basin Water Resources Management Decision Support System Integration Research" (project number: 91325302), training project "Heihe River Basin Blue Green Water Research" (No. 91025009) funding. But also by the National Natural Science Foundation of the project "Beijing-Tianjin-Hebei Water Shortage in the Evolution of the Law and Drive Mechanism" (No. 41571022), Beijing Natural Science Foundation key project" Beijing-Tianjin-Hebei Water Footprint Drive Mechanism and Water Resources Regulation Analysis" (item number: 20140505) and the Central Organization of the first batch of" Young Top-Notch Talent" part of the funding. In the above research, the Chinese Academy of Sciences Cold and Arid Regions Environmental and Engineering Research Institute Cheng Guodong Academician, Chinese Academy of Sciences Ecological Environment Research Center Fu Bojie Academician, China Agricultural University Kang Shaozhong academician, Institute of Geographic Sciences and Resources, Chinese Academy of Sciences researcher Li Xiaobin, Chinese Academy of Sciences Institute of Environment and Engineering, Institute of Environment and Engineering, Chinese Academy of Sciences, Institute of Soil Science, Chinese Academy of Sciences Zhang Guilin researcher gave a lot of guidance and help, hereby sincerely thanked. In addition, we would like to pay special tribute to the National Natural Science Foundation of China, in particular the Director of the Department of Science, Song Changqing, the Director of Cold Shining and the Director of International Cooperation Bureau Zhang Yinglan and Director Zhang Yongtao. The completion of this book has received a lot of support from experts and colleagues, and we are particularly grateful to Academician Cheng Guodong and Fu Bojie for their guidance and help on our work. Also thank Professor Zhang Yinglan, Professor Li Xiubin, Xiao Honglang researcher, Zhang Ganlin researcher in the project implementation and research work in the guidance and help. In addition, we are particularly grateful to the College of Environmental Science and Engineering, School of Environmental Science and Engineering, South China University of Science and Technology, Cold and Arid Regions Environmental and Engineering Research Institute of the Chinese Academy of Sciences, Forestry College of Inner Mongolia Agricultural University, Field Forest Ecosystem Research Station of Daxinganling, Inner Mongolia, Inner Mongolia root River Forestry Bureau and other units of our strong support and help.

Due to the limited level of the author, coupled with the complexity of the water resources research itself, the book is inappropriate, please readers criticized.

The author
March 2016 in Beijing

Contents

Chapter 1 Introduction

1.1 Research background and significance

In nature, all organisms need fresh water resources to ensure their survival (Oki et al., 2006). Ensuring adequate supply of fresh water resources is essential not only for human but also for ecosystems. Water resources in the general sense refers to the freshwater that can be used by the ecological environment and human society in the water cycle. The source of the water supply is mainly precipitation, which is surface water, groundwater and soil water. It can be renewed year by year through water cycle (Cheng Guodong et al., 2006). Renewable freshwater resources are essential natural resources and strategic resources to maintain terrestrial ecosystem health, human food security and ecosystem security. In the twentieth century, the total amount of renewable freshwater resources formed by precipitation remained essentially unchanged, while human water demand surged six times. Thus, there is a situation in which human life and water production and water systems compete for water, which is particularly severe in China (Liu et al., 2013; Xu et al., 2013). Human production and living water squeeze the phenomenon of frequent water use of ecosystems, some of the ecosystem has been seriously degraded. Water can be divided into blue water and green water. Blue water is mainly rivers, lakes and shallow groundwater, green water is derived from precipitation, stored in unsaturated soil and plant absorption and use of transpiration of the part of the water (Falkenmark, 1995; Falkenmark et al., 2003). In recent years, blue-green water research has led to the scientific community on the concept of water resources and evaluation of the rethinking, and gradually affect the human water resources management thinking, has become a hot topic in the field of hydrology and water resources (Cheng Guodong, 2006). Falkenmark (1995) firstly introduced the concept of green water and reviewed the role of green water in terrestrial ecosystems, the progress of green water evaluation studies, factors affecting green water flow, global green water resources and water security. The green water resources into the evaluation of water resources, focusing on blue water and green water resources comprehensive evaluation, natural ecosystems and food production green water balance research, and attention to green water management. The concept of blue-green water has been raised and has received extensive attention from hydrologists, ecologists, agronomists, environmentalists and many international agencies. Comprehensive evaluation of green water flow and blue water flow has also become an important direction for water resources research (Falkenmark et al., 2006; Schuol et al., 2008). Green water flow is defined as the actual evapotranspiration, that is, the flow of water vapor to the atmosphere, including water vapor flow such as farmland, wetland, surface evaporation, vegetation retention and so on. The blue water flow includes surface runoff, soil flow (slope flow) , underground runoff (Schuol et al. ,2008;Zang et al. ,2012). From the perspective of the global water cycle,65% of the total precipitation on the global scale is returned to the atmosphere through forests, grasslands, wetlands and farmland, and only 35% of the precipitation is stored in rivers, lakes and aquifers, Became blue water(Falkenmark, 1995; Cheng Guodong et al. ,2006).

The emergence of the concept of blue water and green water not only broaden the connotation of water resources, but also provides new theories and ideas for water resources management (Cheng Guodong et al. ,2006; Xu Zongxue et al. ,2013). How to quantitatively evaluate blue water and green water resources has become one of the most scientific problems in the field of hydrology and water resources research. In the past, people's understanding of watershed or regional water cycle was mainly blue water, but little was known about the ecosystems and green water, which was extremely important to mankind. In this study, we selected the Heihe River, a typical inland river in arid area, and studied the spatial distribution pattern, evolution law and driving mechanism of blue-green water in the watershed scale with blue-green water as the core. Based on the water footprint concept, Shortages and sustainable use of water resources for the national inland river basin integrated management, water security, ecological security and economic sustainable development to provide theoretical basis and scientific and technological support.

1.2 Research status of blue-green water at home and abroad

Traditional water resources assessment and management focuses on surface and groundwater, is "blue water", ignoring the important water "green water" of ecosystems and rainfed agriculture(Falkenmark, 1995). The concept of blue water and green water is put forward, so that the water cycle and ecological processes are closely linked, reflecting the vegetation and hydrological processes affect the relationship between each other. At the international level, the conceptual system and evaluation method of blue-green water are still in the early stage of development, but the evaluation of blue-green water has been paid more and more attention in the field of hydrology and water resources (Rockström et al. , 2010) . Stockholm International Water Resources Research Center (SIWI) , the United Nations Food and Agriculture Organization (FAO) , the International Water Resources Management Institute (IWMI) , the International Fund for Agricultural Development (IFAD) , the Global Water System Project Team (GWSP) Began to focus on green water research.

At present, the evaluation of green water is mainly concentrated on the global orregional scale, focusing on the evaluation of green water resources and its spatial and temporal distribution(Falkenmark Rockstrom, 2006; Rost et al. ,2008; Liu Y. et al. ,2009; liu J. et al. ,2010, 2013). The On a global scale, the return of green water from the natural ecosystems and farm ecosystems through forests, grasslands, wetlands, etc. to the atmosphere accounted for 61.1% of the total precipitation of terrestrial ecosystems. Less than 40% of the precipitation was stored in rivers, lakes as well as shallow groundwater layers, became blue water (Schiermeier, 2008). Green water has an irreplaceable role in global ecosystems and food production. Liu et al. (2009a) have found that more than 80% of the world's food production is dependent on green water. While grassland and forest ecosystem water supply is mainly dependent on green water. The evolution of blue-green water caused by the change of land use type has also become a hotspot (Gerten et al., 2005; Jewitt et al., 2004; Liu et al., 2009).

At present, the method of estimating the amount of green water resources can be divided into the following three categories:

(1) Utilizing the evapotranspiration required by the dry matter of the main ecosystem to multiply the primary productivity to assess the amount of green water resources. Postel et al. (1996) used net primary productivity data to estimate the evapotranspiration of global rain-fed vegetation (natural forests, grasslands, plantations and rain-fed crops) and obtain the evapotranspiration of the major land-use types, known as green water resources.

(2) The use of hydrological or ecological environment model to assess the green water flow, the water is divided into two parts of blue water and green water. Jewitt et al. (2008) used the Agricultural Watershed Study Unit (ACRU) model and the Hydrological Land Use Change (HYLUC) model to estimate the amount of blue water and green water resources in the nine land use scenarios in the Mutale watershed in southern Africa. Schuol et al. (2008) used the ArcSWAT model and the SWAT-CUP uncertainty analysis algorithm to estimate the amount of blue water green water resources on the African continent. Faramarzi et al. (2009) estimated the effects of different irrigation practices on wheat yield by estimating the amount of blue water and green water resources on a monthly scale in Iran under reservoir operating conditions. Liu et al. (2009a; 2010) used the ecosystem process model (GEPIC) to evaluate the impact of blue-green water consumed by global agricultural production and the management of blue-green water consumption with high spatial resolution. Rost et al. (2008) coupled the hydrological model with biogeography and biogeochemistry to estimate the flow of green water and develop the Global Vegetation Dynamics Model (LPJ). Wang Yujuan et al. (2009) quantitatively simulated the ecological water consumption of vegetation in the Sanmenxia area since the 1950s, and calculated the consumption of green water in different vegetation types. Siebert et al. (2010) used the Global Crop Water Requirement Model (GCWM) to estimate the amount of blue and green water resources required for global crops in 1998 and 2002. Wu Hongtao et al. (2010) used the AvSWAT model to estimate the amount of green water resources in the Biliu River Basin.

(3) Estimate the amount of green water resources based on the actual evapotranspiration and spatial information of typical ecosystems. Rockstroem et al. (2001) used the area of the biota in the woodland, grassland, and wetland to multiply the evapotranspiration, using the product of water use efficiency and crop yield to estimate the green water flow. In the above three methods, the method of model evaluation is paid more and more attention by the international scientific community because of its low cost, easy to carry out large-scale high spatial resolution re-

search, can carry on the scene analysis and so on many reasons.

In the country, blue-green water research has just started, the relevant research is still very scarce. Cheng Guodong and Zhao Wenzhi (2006) took the lead in introducing the concept of green water and its role in terrestrial ecosystems, and advocated our scientists to strengthen green water related research. Liu Changming and Li Yucheng (2006) based on the concept of green water, blue water and broad water resources, clarified the green water and ecosystem water, green water and water-saving agriculture. After the publication of the above literature, the concept of green water has gradually been familiar with domestic scholars, green water evaluation methods and key scientific issues have been gradually elaborated (Li Xiaoyan et al., 2008; Qiu Guoyu, 2008). Recently, our scholars have begun to carry out some exploratory research on blue-green water evaluation. Liu et al. (2009b) applied the GEPIC model to evaluate the blue-green water of the global farmland ecosystem using a spatial resolution of 0.5 radix (approximately 50 km × 50 km per raster), resulting in more than 80% of the global farmland ecosystem (2009b) decomposed the green water flow from Chinese farmland into productive green water (vegetation transpiration) and unproductive green water (soil evaporation), and the results showed that the productive green water in the farmland ecosystem accounts for about two-thirds of the total green water. Wu Hongtao et al. (2009) used the SWAT hydrological model to assess the temporal and spatial distribution of green water in the upper reaches of the Biliu River. Liu et al. (2009) quantified the changes in blue water in the Laohe River Basin in northern China due to land use and cover changes. Wu Jinkui (2005), Cheng Yufei (2007), Tian Hui (2009), Jin Xiaomei (2009), Gao Yangyang (2009), Li et al. (2010) and so on have made use of satellite remote sensing to study the influence of evapotranspiration in the Heihe River Basin. Wen Zhigun et al. (2010) to do a typical vegetation type under the green water cycle simulation. At present, blue-green water assessment at home and abroad is mainly concentrated on the global or regional scale, the accuracy is not high and difficult to apply directly to the actual river basin water management. In the watershed scale, blue and green water resources and water mode integration research is still rare. Moreover, Liu et al. (2009a; 2009b) found that human activities (such as irrigation) had a significant effect on blue-green water evolution.

Although scholars have done some preliminary exploratory studies on blue-green water evolution and land use (Gerten et al., 2005; Liu et al., 2009b; Zhao Wei, 2011). However, it is rare to study the laws and mechanisms of blue-green water evolution in the changing environment. In conclusion, the research on blue-green water can not reveal the climate-hydrology-ecology-human relationship in the watershed scale, and the lack of sufficient scientific basis for the application of blue-green water concept to the river basin water resources management.

1.3 Research status of water resources evaluation in the Heihe River Basin

China is ranked by the United Nations as one of the 13 poor countries, accounting for 1/3 of the northwest arid inland river area due to congenital water shortages, coupled with unreasonable use of water resources, making the water problem has become the local economic development and Key issues of ecological protection(Cheng Guodong et al. ,2006). The impact of human activities on the hydrological cycle and ecosystem of arid areas is quite prominent, and the problem of ecological degradation is also serious. Fully understanding the river basin water cycle is the basis of river basin water management(Xia Jun et al. ,2009). However, the current understanding of the inland river basin water cycle is still based on surface water, taking into account the groundwater, the watershed ecosystems and human beings is extremely important green water resources and water use patterns are poorly understood, the urgent need to form a relatively complete basin scale. The blue-green water comprehensive evaluation of the theoretical framework to support the watershed scientific research.

The Heihe River Basin is a typical arid region inland river basin, with typical characteristics of the inland river basin in arid areas(Cheng Guodong et al. ,2003). The Qilian Mountains to the northern desert areas in the south of the Heihe River Basin form a chain-like oasis-desert complex ecosystem with the water cycle as the link, from high to low, from the southeast to the northwest, showing the typical landscape of the inland river basin in the northwest Pattern characteristics(Cheng Guodong et al. ,2003;2006). As a result of the severe desertification in the middle and lower reaches of the Heihe River, it has become a major source of dust storms and has formed a strong dust storm that has spread to the north of China and even in East Asia. This has aroused great attention from the Chinese government and the wide attention of the community at home and abroad. Cheng Guodong et al. ,2003;2006).

The Heihe River Basin hydrology and water resources research in this context have also made rapid progress. Cheng Guodong et al. (2006) estimated the amount of blue water resources in the Heihe River Basin, such as runoff and glacier meltwater runoff. Chen et al. (2003), Han Jie et al. (2004) used TOPMODEL to simulate the runoff in the upper reaches of the Heihe River. Wang et al. (2003), Huang Qinghua et al. (2004) attempted to apply the distributed hydrological model SWAT model to the upper reaches of the Heihe River Basin, and achieved good simulation results. Jia Yangwen et al. (2006a;2006b) based on the mechanism of water cycle, a distributed simulation model WEP-HeiHe of Heihe River Basin water circulation system was developed on the basis of the influence of artificial water cycle on the basis of ArcGIS, which was applied to the upper reaches of Heihe River Simulation prediction. Li Hongyi and Wang Jian (2008) used the snowmelt runoff model to simulate the runoff during the snow melt in the upper reaches of the Heihe River. Yang Mingjin et al. (2009) analyzed the variation of runoff in the watershed due to the 55 years of natural runoff sequence of the Yingliagan hydrological monitoring station in the Heihe River Basin. Zhao Yingdong et al. (2009) used statistical methods to analyze the variation of air temperature, rainfall, evaporation and runoff in the Heihe River runoff area. Wang et al. (2010) and Guo Qiaoling et al. (2011) analyzed the effects of climate change on surface runoff in the Heihe River Basin using the data of typical weather stations in recent 50 years. Zhou et al. (2009) applied the method of remote sensing, geographic information system and geostatistics to analyze the response of land use change to spatial and temporal variability of groundwater level in the middle reaches of the Heihe River Basin. Zhang Yinghua et al. (2009) studied the groundwater recharge mechanism in the middle reaches of the Heihe River. Zhao Jianzhong et al. (2010) summarized and analyzed the conversion of surface groundwater in the middle reaches of the Heihe River. He Chansheng et al. (2009) applied the distributed large watershed runoff model to assess the impact of climate change on hydrology and the impact of glacier retreat on midstream and downstream waterfloods. Li et al. (2011) demonstrated that the decline coefficient of the Heihe River Basin reached 58% through the degradation of the four-river glacier permafrost in the Northwest, and thus had a significant effect on the hydrological processes of the basin. Wang et al. (2011) used the remote sensing technique and the AIEM model to study the effects of soil water and surface roughness on hydrology in the Heihe River Basin. Zhou et al. (2011) used the remote sensing technique and FEFLOW/MIKE11 model to do the study of groundwater conversion in the middle reaches of the Heihe River. Li et al. (2010) used four climate models to combine the SWAT model with three different scenarios to make the impact of climate change on the Heihe River water resources and to predict hydrological and climate change in the Heihe River Basin from 2010 to 2039. The results show that the runoff of the basin will be −19.8%-37% change, the soil water will have −5.5%-17.2% change, and the evaporation will increase 0.1%-5.9%. Li et al. (2009; 2010) used the SWAT model to simulate the upper reaches of the Heihe River Basin, and analyzed the uncertainty of the model.

In general, in the Heihe River Basin, blue water is still the focus of scholars to study, and for the ecosystem and human beings is extremely important green water research is still scarce. At present, the study of blue-green water can not reveal the climate-hydrology-ecology-human relationship on the scale of river basin, and the lack of sufficient scientific basis for the application of blue-green water concept in river basin water resources management. This paper discusses the spatial distribution pattern of blue-green water in the inland river basin in the arid region, discusses the transformation rule of blue water and green water, discusses the key water cycle and ecological problems of the arid area, discusses the watershed sustainable utilization of water resources and management strategies to achieve sustainable development of the basin has important theoretical and practical significance.

Since the beginning of 2010, the National Natural Science Foundation of China initiated the project to support the "Heihe River Basin ecology-hydrological process integration research" major research project. This paper takes the Heihe River Basin as a typical research area in China, and starts the process of understanding the interaction between the ecosystem and the hydrological system in the inland river basin by establishing the scientific observation-test and data-simulation research platform of the inland river basin in China. And the mechanism, the establishment of watershed ecology-hydrological process model and water resources management decision support system to improve the inland river basin water-eco-economic system evolution of the comprehensive analysis and forecasting ability for the national inland river basin water security, ecological security and economic sustainable development to provide basic theory and technology support. The plan has a total of nearly 100 important projects, key projects and nurturing projects, after hundreds of scholars hard work in the hydrological model development, water resources assessment, groundwater exploration, glacial and frozen soil hydrological processes, riparian forest water monitoring and the protection and precipitation of the source and accounting, ecological-hydrological process monitoring and other aspects of a large number of high levels of research results. "Heihe River Basin Blue Green Water Assessment" is a project funded under the framework of the major research program, with a view to considering water resources and their sustainable use in a more comprehensive manner, taking into account both blue water and green water.

1.4 Hydrological model simulation of blue-green water research progress

In order to study the spatial and temporal evolution of blue-green water on the scale of river basin, it is necessary to simulate the hydrological process of the Heihe River Basin by hydrological model. From the current situation, the semi-distributed hydrological model SWAT (Arnold et al., 2005) developed by the US Department of Agriculture can be used for the simulation and prediction of hydrological cycles in the basin and has been widely used at home and abroad in recent years (Pang Jingpeng et al., 2007; Zhang et al., 2008). SWAT model is one of the most widely used models in the world. It is mainly used to evaluate the distribution of water resources and its changes. It can also identify and simulate key non-point source pollution. The design module is more comprehensive and mature. Better meet the needs of users (Gassman et al., 2010). SWAT model has been widely used in water quality assessment and surface runoff simulation at the basin scale (Hao Fanghua et al., 2006). At present, a large number of domestic studies have used SWAT model to simulate the effects of climatic conditions and land use types on the hydrological cycle of the basin, such as the Haihe River Basin (Wang et al., 2008; Li Jianxin et al., 2010), the Yellow River Basin (Li et al., 2010), the Huaihe River Basin (Bole Lei Lei et al., 2010), the Chaohe River Basin (Guo Junting et al., 2012). The SWAT model can be combined with other models or models of modules to simulate and evaluate the impact of climate change and human activities on hydrological processes in a watershed. Stonefelt et al. (2000) and Fontaine (2001) used SWAT model to couple climate change scenarios to simulate the effects of CO_2 concentration on vegetation growth and runoff. SWAT model has been widely used in river sediment yield, farmland pesticide transfer, watershed and non-point source pollution in irrigation area (Wang Xiaoyan et al., 2009; Zhang Yongyong et al., 2009; Sun Yongliang et al., 2010).

The rate of the input parameters of SWAT model and the process of verification will directly affect the merits of the model simulation results. The accuracy of the evaluation of the SWAT model is usually determined by the coefficient of determination (R^2) and the Nash coefficient

(E_{ns}). A large number of studies at home and abroad have improved the parameter sensitivity and parameter rate of SWAT model, and obtained the ideal results (Huang Qinghua et al., 2010). Gassman et al. (2007) concluded that most of the SWAT model simulations were better than the accuracy of the simulation of the hydrological process when the 37 SWAT models were used to simulate the loss of pollutants. At the same time, the model is not ideal for day scale simulation, mainly because the daily scale of the input data can not fully represent the characteristics of these watersheds. SWAT model in the practice of application process has been improved and improved. From SWAT 94.2 to SWAT 2009 in the 1990s, the SWAT model has undergone multiple changes, and each modification has been refined or added new features. Eckhardt et al. (2002) have studied the calculation formula of osmosis and soil flow in SWAT model for the characteristics of steep slope, thick bedrock, shallow soil and groundwater in runoff in central German mountains, The main flow of the process. Katsanova et al. (1998) developed the SWAT-MOOD model by coupling the MOD-Flow model with the SWAT model and verified it in the Rattlesnake Valley, Kansas, USA. SWS model based on ArcviewGIS platform AvSWAT-2000 and AvSWAT-2005, and ArcGIS platform based ArcSWAT are running in the same structure SWAT model, it has a very powerful spatial analysis and processing functions. In summary, the SWAT model has been developed and applied by scientists at home and abroad for more than 30 years. The practicability and accuracy of the model have been popularized and validated in different watersheds and regions around the world.

International and domestic applications exist specifically for blue-green water evaluation (Faramarzi et al., 2009; Schuol et al., 2008; Wu Hongtao et al., 2009). The SWAT model first divides the watershed into several sub-basins according to the terrain, and then divides the sub-watershed into several hydrologic response units (HRU) according to the soil properties (Wei Huabin et al., 2007) and land use. SWAT is based on the hydrological response unit soil water balance, simulating surface runoff, soil evaporation, vegetation transpiration, shallow and deep groundwater leakage and other hydrological processes. The surface runoff is calculated by the modified runoff curve method (SCS), and the potential evaporation can be calculated according to the data acquisition conditions such as Penman, Penman-Monteith or Hargreaves (Arnold et al., 1998). The calculation of other hydrological processes is detailed in Arnold et al. (2005). The focus of this study is to apply the SWAT model to the Heihe River Basin of the inland river in the arid area, and to analyze the temporal and spatial distribution pattern of blue-green water (including the spatial distribution of blue-green water resources and the analysis of blue-green water characteristics in different typical years). The SWAT model evaluates the blue water flow by the amount of water produced (the amount of water entering the river at a given time) and the sum of underground runoff. The green water flow is evaluated by actual vegetation transpiration and soil evaporation. The amount of water stored in the soil layer over a period of time to evaluate the amount of green water storage. The spatial heterogeneity of terrain, soil, meteorology, land use, management, and water conservancy engineering determines the spatial distribution of blue and green water, and the meteorological parameters of different historical periods determine the difference of blue and green water in different typical years. Based on the above information of the Heihe River Basin, the basic database of the Heihe River Basin is established, and the SWAT model is used as the input data. Based on this, the spatial distribution pattern of blue-green water and blue-green water characteristics in different typical years are analyzed.

1.5 Evaluation method of water shortage

International water shortage assessment methods are usually based on the number of local water resources and water consumption, there are four types of evaluation methods.

The first method for the Falkenmark index method, first proposed by the Swedish scientist Professor Falkenmark (1989). Based on the calculation of per capita water demand, this method sets several thresholds of water shortage: the per capita water resources is less than 1,700 m^3 per year is the shortage of water resources; less than 1,000 m^3 per year is the shortage of water resources; less than 500 m^3 per year for absolute water shortage. This method is easy to understand, the required data is also easy to obtain, so in the early stages of water resources has been widely used. However, this approach only takes into account the amount of water resources in the area and does not reflect the mitigation of water resources shortages, nor does it reflect changes in water demand in different regions due to differences in climate conditions and lifestyles. Ohlsson (1998; 1999) combines the per capita water resources with the Human Development Index of the United Nations Development Program (UNDP) to reflect the resilience of society and is a refinement of the Falkenmark indicator.

The second approach is to compare water resources with water use to assess water shortages. Typical represent the criticality ratio used by Alcamo et al. (2000; 2003), ie. the ratio of water consumption to the total amount of renewable water resources. This method is widely used in water resources evaluation, such as Vörösmarty et al. (2000) published an article in Science, based on this method, using a spatial resolution of 0.5 radian to evaluate global water resources and water shortages. The Oki and Kanae (2006) published an article in Science, using a similar approach to assess the current state of the world's water resources and the future shortage of water resources, and predicted that in the worst case, the future will be 2/3 of the population facing varying degrees water shortage. However, there are still some limitations in this approach, and if there is no consideration of how much water is available to humans under existing water resources, the calculated water use data does not represent water consumption data; pressure and other factors.

The third method proposed by the International Water Resources Management Institute (IWMI), in the evaluation of water shortages to consider the proportion of human water demand and water resources and the actual water supply capacity, water shortage is divided into physical water (physical water scarcity and economic water scarcity (Seckler, 1998). Physical water shortage refers to the local water resources, can not meet and adapt to economic development caused by water stress. Economic water shortage refers to the abundant local water resources, but the need for water conservancy facilities investment and construction, to be able to develop and use water resources. The evaluation method is more complex and difficult to obtain comprehensive data, and this method is only applicable to the national level of evaluation, it is difficult to carry out small-scale water evaluation.

The fourth approach is the water poverty index developed by Sullivan et al. (2003), which reflects both the amount of water resources and the actual water supply capacity, and the ecological effects of water use (Sullivan, 2002). Indicators take into account the availability of water, water resources, water use, water management capacity, environmental impact and other factors. Similarly, this approach is more complex and suitable for national scale evaluations.

In the water resource shortage assessment system, water pollution is rarely a shortage of water quality, that is, water pollution is as an important reference in the evaluation system (Falkenmark et al., 1989; Vörösmarty et al., 2000; Oki Et al., 2006). Water shortage assessment often only focus on the number of shortages, and water quality issues to consider separately, water quality and water quality evaluation method is not yet mature. However, water shortages and water quality deterioration have also become a limiting factor in the sustainable development of many regions and countries (Vörösmarty et al., 2000; Oki et al., 2006; José et al., 2010; Kummu et al., 2010). For example, China is listed by the United Nations as one of the 13 poor countries, with a shortage of congenital water resources. With the population growth and socio-economic development, water resources are subject to various pollution, resulting in deterioration of water quality. Water pollution caused by water quality and water shortage of resources affect each other, so that China's water shortage situation is more serious. A simple assessment of water quality and water shortage assessment without water quality does not identify the impact of water pollution on the amount of available water resources. China's traditional water pollution assessment indicators are often divided by different types of water quality, the State Environmental Protection Administration (now the People's Republic of China Ministry of Environmental Protection) promulgated the "surface water environmental quality standards" as a basis for assessing the status of water quality. In the water resources bulletin, the most common method is to evaluate the length of the river, lake area, the number of reservoirs, and the number of groundwater monitoring wells that meet different water quality standards. This method can evaluate the water quality of the water, but can not reflect the impact of different human activities on the deterioration of wetland water quality. The aquatic eco-environmental carrying capacity is another indicator of water pollution capacity, which aims to quantify the maximum discharge load or the amount of pollutants that the water body can accommodate under certain environmental and hydrological conditions. Environmental carrying capacity can reflect the wetland water purification capacity, but can not quantify the interpretation of a certain amount of sewage on the impact of water purification capacity. In addition, China's traditional water pollution assessment methods are integrated pollution index method, fuzzy mathematics method, artificial neural network analysis method and thermodynamic methods(Fu Guowei et al. ,1985;Huang Zhihong et al. , 2005;Li Xianghu et al. Zhang et al. ,2003). Similarly,these methods are mainly to evaluate the degree of contamination of contaminated water bodies, and there is little research on the relationship between the quantity and quality of water resources.

The international community increasingly pay attention to the combination of water quality and water research, our work in this area has gradually started. For example, Xia Jun et al. (2006) constructed a joint evaluation method for water quality and water quality to evaluate surface water use. The main characteristic of the method is to evaluate the distribution of water quality in the total water resources based on the correspondence between water quality and water quality processes of the unit and the composite system. Xia Xinghui et al. (2005) established a joint water quality and water quality evaluation method and applied it to the Yellow River Basin, and put forward the concept of water resources functional capacity and water resources deficit. The study of the river's natural runoff is divided into three categories, from the river for industrial, agricultural and living water, flow to the ocean or other watershed waters, stored in the river water. If the actual water quality is higher than the water quality requirements of water resources, it indicates that the water body can meet the requirements of higher water resources and have the function of water resources. If the actual water quality is lower than the water quality requirement, resource function deficit. The evaluation method based on the socio-economic system on the needs of water resources, the total amount of water resources can be assessed by the amount of water, it is difficult to reflect the river ecological water needs. Wang Xiqin et al. (2006) set up a comprehensive evaluation method of ecological water demand from the natural and social water cycle, as well as the river water quality(such as pollution ratio), and applied in the Liaohe River Basin. Zhang Yongyong et al. (2009) Based on the SWAT model and the distributed water quality coupling model, a joint water quality evaluation method based on water cycle process is proposed and applied to the Haihe River, in this paper, watershed area. However, these methods are mainly used in the watershed scale, and have not yet been formed to be applicable to multi-scale water shortage assessment of the general method. Although the domestic and international water quality and water quality evaluation has made great progress, the current assessment of water shortages is still based on the amount of blue water resources, water quality water shortage is often stuck in qualitative evaluation, can not meet the consideration of water and water quality Of the comprehensive lack of quantitative assessment of water needs.

1.6 Evaluation of water shortage based on water footprint

The water footprint is a concept of water resource consumption indicators proposed by Professor Hoekstra et al. Of the Tuen Tuen University in Turkey(Hoekstra et al. ,2003), defined as the total amount of freshwater consumed in the process of human production and consumption, including direct water(Hoekstra and Hung,2002). Water Footprint Assessment, as an analytical tool, it is a good way to describe the close relationship between human activities and water shortages, as well as a comprehensive water resource management. Innovative approach (Hoekstra et al., 2011), it is a multidimensional indicator of the amount of water consumed, the type of water source and the type of pollution and pollution, and all its components define the time and place of water consumption and pollution.

Watcr footprint is divided into blue water footprint, green water footprint and gray water footprint. Blue water footprint refers to the consumption of blue water resources, ie. surface water and groundwater, that are consumed in the production of products or services. Green water footprint refers to the green water resources consumed in the human production process. Gray water refers to the water that absorbs contaminants during human production (Hoekstra et al., 2011). The concept of the gray water footprint was first proposed by Hoekstra and Chapagain (2008) in 2008, and the continuous improvement of the gray water footprint team (Zarate et al., 2010) through the water footprint network is defined as the natural background and the existing water quality Environmental standards as a benchmark, will be a certain amount of pollution load absorption assimilation required fresh water volume. The gray water footprint achieves the purpose of assessing the degree of water pollution from the point of view of water capacity and can more directly reflect the effect of water pollution on the amount of available water resources.

The study of water footprint at home and abroad mainly focuses on the following five levels: process, product, department, administrative region and global. At the process level, Chapagain et al. (2006) calculated the water footprint of cotton in different production processes. At the product level, Mekonnen and Hoekstra (2011) evaluated the blue, green and gray water footprints of 126 crops worldwide during 1996-2005; products such as pizza (Aldaya and Hoekstra, 2009), coffee and tea (Chapagain and Hoekstra, 2007). Of the water footprint also have scholars to account. At the sectoral level, Aldaya et al. (2010) calculated the water footprint of the Spanish agricultural, industrial and living sectors and found that the main reason for the shortage of water resources in Spain was the irrational distribution and management of water resources in the agricultural sector. At the national level, China (Liu and Savenije, 2008; Ma et al., 2006), India (Kampman et al., 2008), Indonesia (Bulsink et al., 2010), the Netherlands (Van Oel et al., 2009), Chapagain (Orr, 2008) and France (Ercin et al., 2012) and other countries also have relevant water footprint evaluation research. At the global level, Hoekstra and Chapagain (2007), Hoekstra and Mekonnen (2012a) have accounted for and evaluated the water footprint of products and services consumed by human activities. Although more and more scholars are conducting research on water footprints, there are relatively few studies on water footprints at the basin level due to the shortage of statistical data at the basin level (UNEP, 2011; Zhao et al. 2010), especially in arid and semi-arid areas. It is necessary to evaluate the water footprint at the basin level, which is not only an important step and process to understand the impact of human activities on natural water cycles, but also a basis for integrated water resources management and efficient use.

In the water footprint method, the gray water footprint provides a new method for quantitative evaluation of water quality shortage. At the international level, the conceptual system and evaluation method of gray water footprint are still in the early stage of development. However, the concept of gray water footprint has been paid much attention in the field of hydrology, water resources and environmental science, and it has caused the international water footprint network, the United Nations Environment Program, The United Nations Food and Agriculture Organization and other international organizations. At present, the study of gray water footprint at home and abroad is mainly focused on two aspects, one is to evaluate the gray water footprint of agricultural products. Such as Chapagain and Hoekstra(2007), calculated the blue water, green water and gray water footprints of global rice from 2000 to 2004, and evaluated them from the perspective of production and consumption. The results show that the reduction of gray water footprint is mainly due to the reduction of field crops Fertilizer and pesticide application, to improve the effective use of water resources. The second is to evaluate the gray water footprint of industrial products. Such as Ercin et al. (2011) studied a water footprint of hypothetical sugary carbonated beverages in 0.5 liter PET (polyethylene terephthalate) bottles, and found that the main influencing water footprint of the hypothetical product The factor is the product's blue water footprint and gray water footprint. As the raw materials of the product (such as sugar, caffeine, etc.) in the production process will also be due to the use of chemical fertilizers and pesticides caused by water pollution, so the choice of raw materials and production methods of the difference, resulting in the final product of gray water Footprints. Chinese scholars in the gray water footprint has begun to try some exploratory research. Gai Liqiang(2010) draws on the concept of gray water footprint to calculate and evaluate the gray water footprint of wheat and maize in North China Plain and suggest that in assessing the utilization of water resources in a country and region, it is necessary to consider fertilizer and pesticide on the impact of water pollution. He Hao et al. (2010) studied the water footprint changes of rice in Hunan Province from 1960 to 2008, and the results showed that the proportion of rice ash water footprint showed a rising trend and pollution was increasing. At present, the study of gray water footprint in China is still at the initial stage of learning and imitating foreign advanced concepts and calculation methods. The research on gray water footprint has not yet formed a complete scientific system, and it is not necessary to guide the rational use of water resources continuous management of the conditions. On the whole, there is no study on the shortage of water quality caused by water pollution in regional and river scale, based on gray water footprint and quantitative evaluation.

Water footprint as an analysis and evaluation tool, can be a good human activities and fresh water resources linked to quantitative assessment of human water consumption, water demand and water pollution. However, the water footprint as a new concept was introduced into the water shortage assessment system is very short time, the use of water footprint assessment tools to a comprehensive analysis of the local water shortage and water quality shortage of water resources research is still very small.

1.7 Research content and technical route

1.7.1 Research content

This paper puts forward the theoretical framework and methods of quantitative evaluation of blue and green water resources, and forms a comprehensive evaluation of water resources shortage in combination with water quality and water quantity. The paper puts forward the theoretical framework and methods of quantitative evaluation of blue-green water resources, The water quality and water quantity of the watershed are evaluated based on the water footprint concept and the water quality of the river basin, and the water quality and water quantity of the watershed are analyzed by combining the water and water integration model and evaluating the spatial distribution pattern of blue-green water in the basin. And to analyze the sustainability of water use in the river basin. A typical study on the evaluation of blue-green water and water shortage in the study area was conducted. This study has developed the theory and method of blue-green water resources evaluation in changing environment, and provided scientific basis for the comprehensive response of water shortage in Heihe River Basin, including the following 45 aspects of the study.

(1) Spatial and temporal dynamic distribution of blue-green water in the Heihe River Basin under natural conditions.

The spatial and temporal distribution pattern of blue-green water in the watershed was simulated by using the semi-distributed hydrological model to simulate the blue-green water formation process under natural conditions (no human activity disturbance) in the watershed scale.

(2) Spatial and temporal dynamic distribution of blue-green water in the Heihe River Basin under the influence of human activities.

The effects of climate change, land use change and irrigation on the spatial and temporal distribution pattern of blue-green water in the river basin were analyzed. The changes of blue-green water in the Heihe River Basin under different scenarios were quantitatively evaluated, and the law of blue-green water conversion in the changing environment was discussed.

(3) Spatial and temporal differences of blue water and green water in the Heihe River Basin in typical years.

The temporal and spatial dynamic distribution characteristics of blue-green water in different typical years (drought year, wet year and plain water year) in the Heihe River Basin were studied, and the variation rule of blue-green water in the Heihe River Basin in different typical years was discussed.

(4) Analysis of the historical evolution of blue-green water in the Heihe River Basin.

Based on the model simulation results and statistical test methods, the historical trend of blue-green water in the basin, upper and lower reaches and sub-watersheds was analyzed, and the future trend of blue-green water was predicted.

(5) Evaluation and sustainable analysis of water resources shortage in the Heihe River Ba-

sin.

Based on the concept of water footprint, the blue water footprint, green water footprint and gray water footprint of the river basin were calculated to evaluate the water quality and water quantity of the river basin, and the sustainability of water resources utilization in the Heihe River Basin was analyzed on the monthly scale.

1.7.2 Technical route

The specific technical route of this book is shown in Figure 1-1. A typical Heihe River in the arid area was selected as the research object, and the basic data such as meteorological, soil, permafrost distribution, land use, runoff, socio-economy and pollutant discharge were collected. The hydrological model of the basin was used to simulate the hydrological process of the river basin, and the model parameters were calibrated by comparing the simulated runoff with the measured runoff and adjusting the model parameters. The hydrological model was used to evaluate the formation process and spatial distribution pattern of blue-green water in the basin. The characteristics of blue-green water in dry, wet and plain period were analyzed. Com-

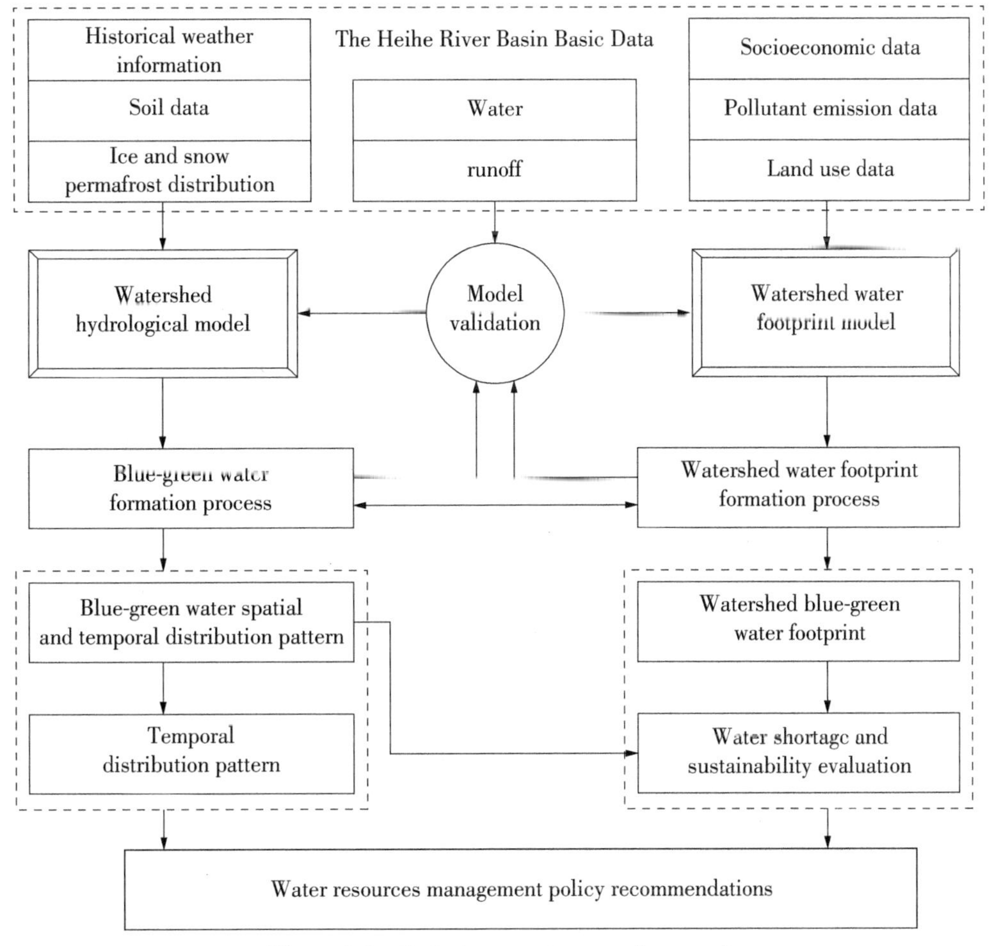

Figure 1-1 Technique route map of research

bined with the natural, hydrological, social and economic situation of the Heihe River Basin, perfect the water footprint model of the river basin, simulate the water and water consumption of the blue and green water in the river basin, calculate the blue water, green water and gray water footprint of the river basin, evaluate the water quality and water quantity, The sustainable use of the Heihe River Basin water resources management policy recommendations for the decision-making departments to carry out scientific management of river basin water resources to provide a theoretical basis.

Chapter 2 Overview of Research Areas and Research Methods

2.1 Overview of the study area

2.1.1 Climate hydrological characteristics

The Heihe River is the second largest inland river in china, located in the northwest inland areas, originated in the northern Qilian Mountains, flows through Qinghai, Gansu, Inner Mongolia three places. The total length of the Heihe River is 821 km, and the upper reaches originate from the Qilian Mountains. The lower reaches of the Yangtze River is finally owned by the People's Republic of China (Ministry of Water Resources, 2002). The Heihe River Basin is located in the middle of the Eurasian continent, far from the ocean, and the river basin climate is mainly affected by the mid-high latitude westerly circulation control and polar cold air mass. The watershed is dry and sparsely populated and concentrated, with strong solar radiation and large temperature difference between day and night. Due to the continental climate and the Qinghai-Tibet Plateau Qilian Mountains-Qinghai Lake climate zone, the middle and lower reaches of the plains and the Alxa Plateau is a temperate zone (Li Zhanling, 2009). The spatial and temporal distribution of water resources in the Heihe River Basin is uneven. Due to the influence of the monsoon, the precipitation changes greatly during the year, and the precipitation in the flood season is large and concentrated. Flood season from June to September, is the largest continuous precipitation for 4 months, the precipitation accounted for more than 73.3% of annual precipitation; winter 3 months (December to February next year) precipitation accounted for 3.5% of annual precipitation. The spring and summer of 5 to 7 months, is the peak of agricultural water, but the precipitation is generally less than normal (Li Zhanling, 2009).

The Heihe River Basin has obvious differences and north-south differences, and the spatial distribution of precipitation is not uniform. In the southern Qilian Mountains, the precipitation decreases from east to west, and the height of the snow line increases from east to west. The precipitation in the central corridor plain area is reduced from the east of the 250 mm to the west down to 50 mm below, and the potential evaporation increases from east to west, from 2,000 mm to 4,000 mm. The average annual temperature of the 2,600-3,200 m area in the Qilian Mountains of the south is 1.5-2.0 ℃, the annual precipitation is more than 380 mm, the maximum is 700 mm and the relative humidity is about 60% (Wang Jinye et al., 2006). Central Hexi Corridor plains rich in light and heat resources, the annual average temperature of 2.8-7.6 ℃, sunshine time up to 3,000-4,000 h, is the ideal area for the development of agriculture (Liu Shaoyu, 2008; Liu Yanyan et al., 2009). In the southern mountainous area, the precipitation increases by 100 m and the precipitation increases by 15. 5-16. 4 mm. For every 100 m above sea level, the precipitation increases by 3. 5-4. 8 mm and the evaporation decreases by 25-32 mm. The lower reaches of the Ejin plains inland hinterland, is a typical continental climate. According to the modern surface water and groundwater hydraulic connection, the Heihe River Basin can be divided into three sub-water systems of east, middle and west. Which is the western river system for the flood river, to discuss the river system; the middle of the Ma Yinghe, Fengle River; the eastern part of the river system for the Heihe River, Liyuan River and the East from the mountain porcelain kiln, Several small rivers (Ministry of Water Resources of the People's Republic of China, 2002). The total runoff from the mountainous area is 3. 755 billion m^3, of which the runoff of the eastern part of the water is 2. 475 billion m^3, including 1. 58 billion m^3 of runoff and the runoff of the Liyuan River. The runoff of the Liyuan River is 237 million m^3, and the other tributary is 658 million m^3 Li Wanshou et al. , 2008).

The source of the Heihe River Basin is about 100 km^2, with an estimated ice storage of 2. 75 billion m^3 and an average annual glacier meltwater of 100 million m^3, accounting for about 4% of the natural runoff of the river. The remaining 96% of the runoff is supplied by precipitation (Cheng Guodong et al. , 2003). The distribution of surface runoff in mountainous area is basically the same as that in precipitation process and high temperature season. In spring, snow and water are the main components in summer and autumn. In late spring and early summer, with the temperature rise, glacier melting and river snow melting, surface runoff increased to May spring flood season, to June runoff accounted for 24. 55% of the total annual runoff, the rainy season (7-9) precipitation increased, glaciers melt water, surface runoff of 55. 71% (Li Wanshou, 2008). In June each year, the river began to increase, June to September appeared summer floods, September irrigation return to water and groundwater overflow, the formation of the peak year of the river; October winter irrigation and precipitation reduction, river runoff again reduced to November to the lowest value, from December to March the following year for non-agricultural water season (Li Wanshou et al. , 2008). In general, the water resources in the Heihe River Basin have the characteristics of river runoff formation, utilization and disappearance, and the river runoff is mainly composed of precipitation. The interannual variation of river runoff is small. The distribution of runoff is concentrated during the year. Groundwater conversion frequently five characteristics (Li Wanshou, 2008; Li Zhanling, 2009).

2. 1. 2 Soil vegetation status

The upper reaches of the Qilian Mountains are affected by mountain climate, topography and vegetation. The soil has obvious vertical band spectrum. The main soils are desert soil, alpine meadow soil (frozen fruiting soil), alpine shrub meadow soil (peat soil type cold frozen soil), subalpine meadow soil (cold carp soil), subalpine grassland soil (cold soil), gray cinnamon soil, mountain chernozem soil, mountain chestnut soil, mountain lime soil, etc. (Liu Hu et al. , 2011). In the middle and lower reaches of the basin, it belongs to gray-brown desert soil and gray desert soil distribution area. In addition to these zonal soils, there are irrigated soils (oasis irrigated tillage soil), salt soil, fluvo-aquic soil(meadow soil), latent soil(marsh soil) and sandy soil. In the downstream Ejina Banner, gray-brown desert soil is the main zonal soil, which is affected by water and salt transport conditions and climate and vegetation. It also distributes sulfate salty soil, forested meadow soil and salinized forest soil, alkaline soil, meadow saline soil, sandy soil and cracked soil and other non-regional soil(Liu Hu et al., 2011). The upper reaches of the Qilian Mountains vegetation is temperate mountain forest grassland, the growth of shrubs and trees, vertical band spectrum is extremely obvious. The mountainous area is 3,900-4,200 m above sea level, and the alpine meadow vegetation zone is distributed at an altitude of 3,600-3,900 m. The alpine shrub meadow zone, the slope is distributed at an altitude of 3,400-3,900 m, and the shady slope is 3,300-3,800 m; the mountain forest steppe zone, the sunny slope is 2,500-3,400 m and the shady slope is 2,400-3,400 m(Ding Songshuang et al., 2010). This vegetation zone has a very important role in forming runoff, regulating river water and preserving water source. The middle and lower reaches of the zonal vegetation for temperate shrubs, semi-shrub desert vegetation(Ding Song shuang, 2010; Su Peixi et al., 2010).

2.1.3 Basin socio-economic and water use status

The population of the Heihe River Bosin is 163.2 million in 2007, of which agricultural population of 112.75 million; 415.93 million mu of arable land, farmland irrigation area of 306.54 million mu, forest and grass irrigation area of 85.55 million mu(Xiao Shengchun et al., 2011). The upper reaches of the basin includes the majority of Qilian County in Qinghai Province and some parts of Sunan County, Gansu Province, with a total population of 59,800 people, 769,000 mu of arable land, 606 mu of irrigated area of farmland, 2.70 million mu of irrigated area of forest and grass. The middle reaches of the region, includs Gansu Province, Shandan, minle, Zhangye, Linze, high Taiwan and other counties(cities), irrigated agricultural economic zone, population 121.20 million, 390.87 million mu of arable land, farmland irrigation area of 289.38 million mu, 44.95 million mu. The lower reaches of the province include Jinta County in Gansu Province and Inner Mongolia Autonomous Region Ejinaqi, population of 66,300 people, 14.37 million mu of arable land, farmland irrigation area of 11.10 million mu, forest and grass irrigation area of 37.90 million mu(Xiao Shengchun et al., 2011). In the middle reaches of the Heihe River Basin, there are 192 main tunnels in the area of Zhangye, Jiuquan and Jiayuguan. The total length is 2,545 km and the average lining rate is 57.5%. According to incomplete statistics, the Heihe River Basin existing electromechanical wells 6,484, the annual mining capacity of 511.5 million m^3(Liu Shaoyu et al., 2008; Li Zhanling, 2009).

2.2 Research methods

2.2.1 Introduction to concepts

Water can be divided into blue water and green water. Blue water is mainly rivers, lakes and shallow groundwater, green water is derived from precipitation, stored in unsaturated soil and the plant in the form of evapotranspiration absorbed part of the water(Falkenmark, 1995;

Falkenmark et al. ,2006;Schuol et al. ,2008). The green water flow is the actual evapotranspiration,ie. the flow of water vapor to the atmosphere,including farmland,wetland,surface evaporation,vegetation retention,etc. water flow;blue water flow includes surface runoff,soil flow(lateral flow),underground runoff three parts(Schuol et al. ,2008;Zang et al. ,2012). At the same time,many studies used the concept of green water(Liu et al. ,2009b),which is the ratio of green watcr(actual evapotranspiration)to total blue-green water. For the purpose of image interpretation of the area of blue water and green water,the paper introduces the concept of blue water depth and green water depth,blue water depth refers to the blue water per unit area;green water depth refers to the green water per unit area.

2.2.2 Data collection and processing methods

2.2.2.1 Hydrological and meteorological data collection

In this paper,the hydrological data of the two hydrological stations from 1980 to 2004,and the weather data of 19 meteorological stations(see Figure 2-1)from 1958 to 2010 were collected,including daily average rainfall,daily maximum temperature,daily minimum temperature,average temperature),daily relative humidity and wind speed. Data from the Natural Science Foundation of the major research program Heihe data group.

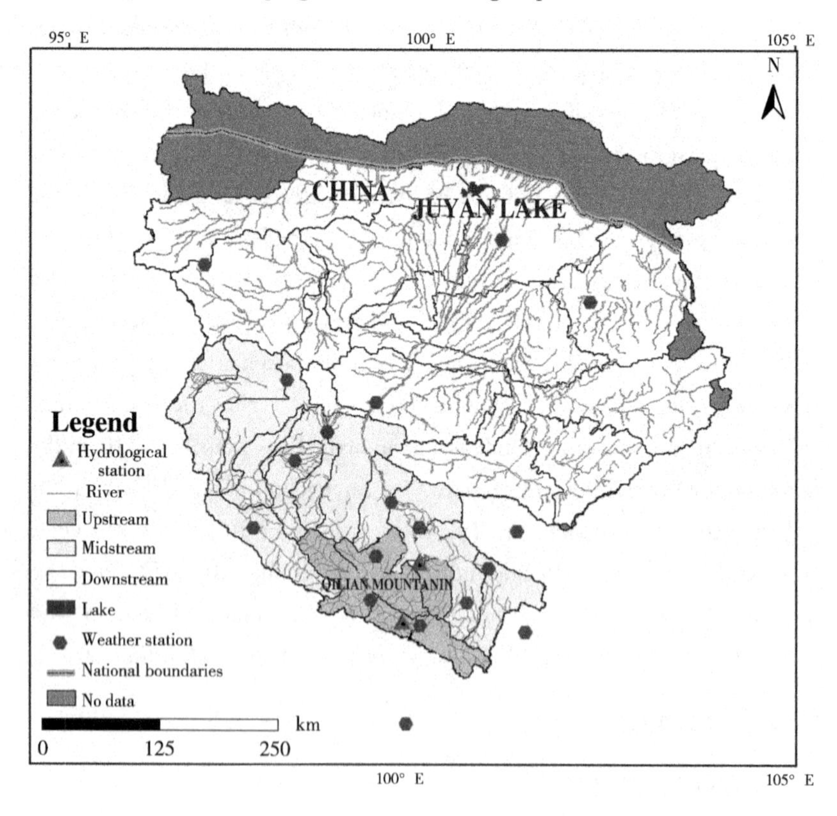

Figure 2-1 Location of the Heihe River Basin,the upstream,midstream and downstream and distribution of weather stations within the basin

2.2.2.2 DEM

In this study, the Chinese Academy of Sciences International scientific data service platform and data sets to provide the spatial resolution of 30 m digital elevation. And cut according to the watershed boundary to obtain the digital elevation of the study area (see Figure 2-2), and then generate watershed slope, slope and other landform index.

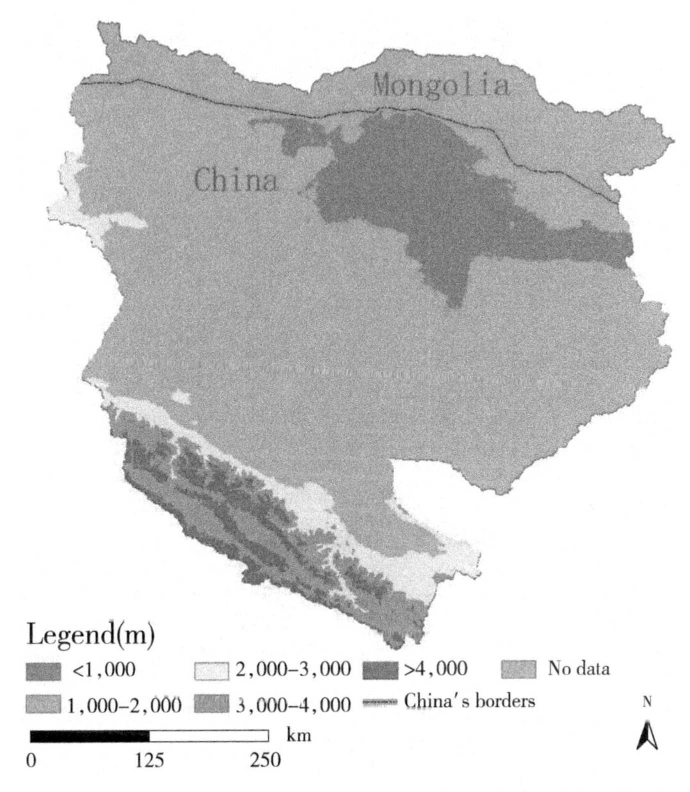

Figure 2-2 The digital elevation map of the Heihe River Basin

2.2.2.3 Land use data acquisition

The land use data of the Heihe River Basin are shown in 1986, 2000 and 2005, with the resolution of 1 km (provided by the Heihe data group). With reference to the latest full-scale land classification standard and the land of the Chinese Academy of Sciences, the land use is divided into forest land, shrub land, grassland, arable land, water area, residential area and construction land and unused land, etc., using the classification data and the data of land type inventory in the Heihe River Basin, a total of 25 types of utilization (Figure 2-3).

2.2.2.4 Soil data

The 1∶1 million soil data used in this study was downloaded from the Harmonized World Soil Database (http://www.iiasa.ac.at/Research/LUC/External-World-soil-database/). The data are provided by the World Food and Agriculture Organization (FAO), the Austrian Institute for International Applied Systems Analysis (IIASA) and the Institute of Soil Science of the Chinese Academy of Sciences (ISSCAS). The database consists of more than 5,000 soil types, and

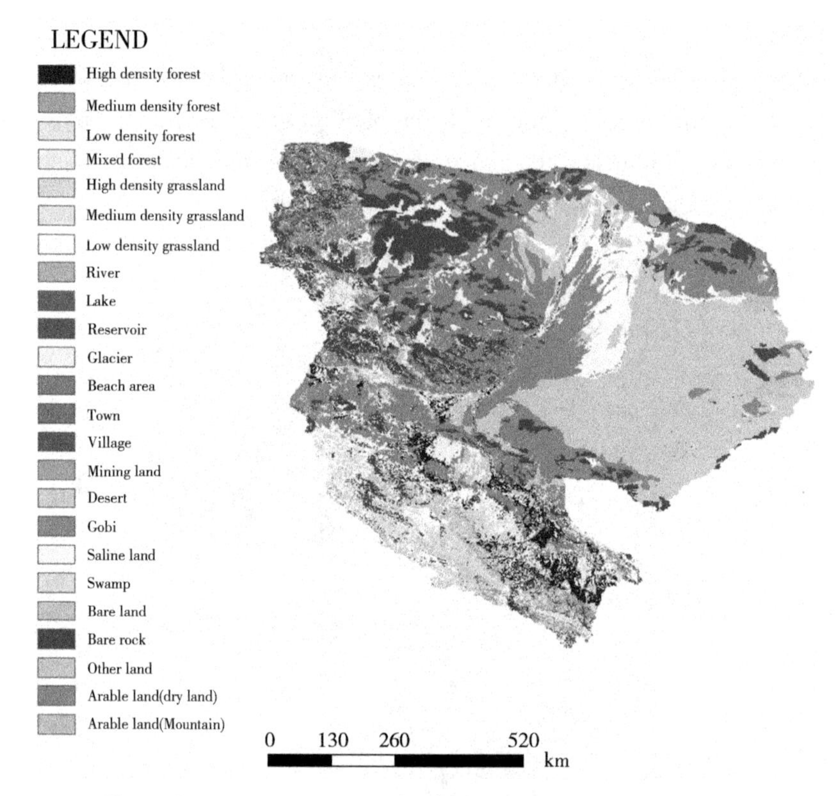

Figure 2-3 The land use types of 2000 in the Heihe River Basin

the soil is divided into two layers(0-30 cm and 30-100 cm). The soil database uses the FAO soil classification standard, in order to meet the needs of model simulation, the use of the process to refer to the US classification criteria, the soil properties to be recalculated. There are 63 major soil types in the Heihe River Basin, among which there are cold soil, alpine meadow soil(frozen felting soil), alpine shrub meadow soil(peat soil frozen felted soil), alpine grassland (cold soil), subalpine meadow soil(cold carp soil), subalpine grassland soil(cold soil), gray cinnamon soil, mountain chernozem, mountain chestnut soil, mountain lime soil, etc. ; the area is mainly gray-brown desert soil and gray desert soil.

2.2.3 SWAT model simulation method

Hydrological model gradually evolved from the traditional lumped hydrological model to a hydrological model with a certain physical mechanism; and from the whole basin as a whole, gradually began to consider the hydrological phenomenon or the spatial distribution of elements, water cycle system elements of the process of interrelated(Li Zhanling, 2009) from the simple analysis of the law of the change of production and distribution in a certain basin, to the temporal and spatial evolution of the influencing environment, and the impact of human activities on hydrological cycles(Li Zhanling, 2009). The hydrological model studies the basin or regional hydrological system by simulating the complex hydrological phenomena in nature and has been developed and perfected in the process of studying and solving the multi-objective decision-making and management of water resources in the basin (Xu et al. ,2010). The hydrological model of the basin can be divided into three types: system theory model, conceptual model and physical model (Xu Zongxue et al. ,2010). This study mainly uses the semi-distributed hydrological model SWAT to simulate. The main reason for choosing the SWAT model is twofold: firstly, the model has successfully conducted water and water quality modeling studies in many countries and regions in the world (Schuol et al. ,2008; Gerten et al. ,2005; Faramazi et al. , 2009; Gassman et al. ,2007). Secondly, the model has been successfully studied in some areas of the upper reaches of the Heihe River, and the results are more ideal (Huang Qinghua et al. , 2004; Li Zhanling, 2009).

2.2.3.1 Model establishment and parameter sensitivity analysis

SWAT model can simulate a variety of hydrological and physical processes in the basin. Due to the spatial and temporal differences between the underlying surface and climatic factors. Therefore, the SWAT model usually first subdivides the watershed into several sub-basins according to the DEM. The size of the sub-watersheds can be adjusted by defining the minimum catchment area required to form the river, and can be further adjusted by increasing the sub-basin exports, and then divided into hydrological response units (HRUs) in each sub-basin. The hydrological response unit is a collection of land surface areas that have the same vegetation cover, soil type and terrain conditions in the sub-watershed. There are two ways to divide the HRU into a sub-basin: one is to choose a combination of land use and soil types that are the largest, as a sub-basin is a HRU; the other is to divide the sub-watershed into multiple a combination of different land use and soil types, ie. , multiple HRUs (Li Zhanling, 2009). The water balance within each HRU is calculated based on precipitation, surface runoff, evapotranspiration, soil flow, infiltration, groundwater reflow, and river migration losses. The surface runoff estimation is generally based on the SCS runoff curve (Arnold et al. ,1998). The permeation uses the storage calculation method, combined with the fracture flow model to predict the flow through each soil layer. Once the water penetrates below the bottom of the root zone, produce reflow. River flow calculations are based on variable storage coefficient methods or Muskingum algorithms (Arnold et al. ,1998). The SWAT model provides three methods for estimating potential evapotranspiration, Hargreaves (Hargreaves et al. ,1985), Priestley-Taylor (Arnold et al. , 1998) and Penman-Monteith method (Neitsch et al. ,2002). In the second HRU division, the minimum threshold of land use and soil area is set to 10, ie. , if the area ratio of a land use and soil type in the sub-basin is less than the threshold, it is not considered in the simulation The area of land use and soil type is re-scaled to ensure that the entire sub-basin area is 100% simulated. In this study, the SCS runoff curve is used to estimate the surface runoff. The flow calculation is based on the Muskingum method (Arnold et al. ,1998). Hargreaves method is used to estimate the evapotranspiration. In this paper, the flow data of two hydrological stations in Zhamushi station and Yingluo canyon station are output and compared with the measured data of two stations, and the applicability of SWAT model in runoff simulation is analyzed. Select the hydrological and meteorological modules in the model, simulate the annual runoff and monthly runoff and evapotranspiration.

The SWAT model contains many parameters, and some parameters can be obtained directly from the input data of the model, such as DEM, soil, land use pattern and its attributes, and some parameters need to be determined by the model. In order to reduce the uncertainty problem in the parameter estimation, it is necessary to analyze the sensitivity of the parameters first, then take into account the physical meaning of the parameters and the related literature, select the parameters that are determined; sensitive parameters can be based on the relevant research and model manual empirical analysis of these parameters, as long as the given experience value is not a great deviation from the true value of the parameters, generally does not significantly affect the simulation results.

2.2.3.2 Rate and validation of models

The rate and validation of the model is a challenging task for any model that really reflects the actual situation. The Nash coefficient (E_{ns}, Nash-Sutcliffe coefficient) and the decision coefficient (R^2, Coefficient of determination; Nash and Sutcliffe, 1970) are used to determine the model and the validation index.

The following formula:

$$E_{ns} = 1 - \frac{\sum_{i=1}^{n}(Q_{oi} - Q_{mi})^2}{\sum_{i=1}^{n}(Q_{oi} - \overline{Q}_o)^2} \tag{2-1}$$

$$R^2 = \frac{\left[\sum_{i=1}^{n}(Q_{oi} - \overline{Q}_o)(Q_{mi} - \overline{Q}_o\right]^2}{\sum_{i=1}^{n}(Q_{oi} - \overline{Q}_o)^2(Q_{mi} - \overline{Q}_m)^2} \tag{2-2}$$

Where: E_{ns} is the Nash-Sutcliffe coefficient, which is used to measure the fit between the model simulation value and the observed value, the closer the value is to 1, the closer the simulation process is to the observed value; R^2 is the correlation coefficient square, R^2 can evaluate the measured value and the simulated data between the degree of fit, When $R^2 = 1$, that is very consistent, when $R^2 < 1$, the value of the closer to 1 indicates that the higher the degree of data consistency; Q_{oi} is the i-th actual observation flow; Q_{mi} is the i-th simulated flow; n is the simulated flow sequence length.

If $E_{ns} < 0$, it is shown that the simulation value of the model is lower than that of the measured value, and $E_{ns} > 0.5$ is used as the evaluation criterion of runoff simulation efficiency. R^2 is closer to 1, indicates that the closer the simulated runoff is to the measured runoff (the higher the explanatory runoff of the simulated runoff to the measured runoff), usually taking $R^2 > 0.6$ as the evaluation criterion for the correlation between the simulated value of the runoff and the measured value (Nash et al., 1970; Yuan Junying et al., 2010).

2.2.4 Statistical analysis method

The statistical method is a very rigorous tool that can be used to test the trend of climate and hydrological data for a long period of time in a region or basin. Routine methods include runoff curve, regression analysis, spectral analysis, parametric test and nonparametric test wait. There are many ways to test the trend of time series, include parameter verification methods such as simple linear fitting method, piecewise linear fitting method (Shao et al., 2002; Shao et al., 2009), nonlinear fitting method, such as polynomial fitting, spline fitting, etc. (Henderson, 2006); also include nonparametric tests such as the Mann-Kendall (M-K) nonparametric statistical test (Burn et al., 2002; Xu et al., 2008). In this paper, M-K statistical test method was used to study the blue water flow and green water flow in the Heihe River Basin from 51 years (1960-2010) We may check the blue and green water, precipitation and temperature to change points with the method of Sequential Version Mann-Kendall (S-M-K). Finally, the Hurst index was used to analyze and predict the future trend of blue-green water and precipitation temperature in the Heihe River Basin.

2.2.4.1 M-K nonparametric statistical test

The M-K nonparametric statistical test is a method recommended by the World Meteorological Organization for time series trend analysis of environmental data and has been widely used to test the trend components of hydrometeorological data, includes water quality, flow, temperature and rainfall (Burn et al., 2002; Xu et al., 2007). The nonparametric statistical method is based on the rank of the data sequence rather than the actual data value to determine the degree of correlation between the two variables, thus avoiding the hydrological research in the large and small values on the impact of the results can be more objective to determine whether the data sequence has a tendency to change (Liu et al., 2002). Assuming $x_1, x_2, \cdots, x_n$ is the time series variable, n is the length of the time series, the M-K method defines the statistic S, and the statistic S of the Mann-Kendall test is calculated using the following equation:

$$S = \sum_{k=1}^{n-1} \sum_{j=k+1}^{n} \mathrm{sgn}(x_j - x_k) \qquad (2\text{-}3)$$

Where, x_j and x_k are the observed values of j th and k th respectively, $j > k$; n is the record length of the sequence, $\mathrm{sgn}(x_j - x_k)$ characterizes the following functions:

$$\mathrm{sgn}(x_j - x_k) = \begin{cases} 1 & \text{if} \quad x_j - x_k > 0 \\ 0 & \text{if} \quad x_j - x_k = 0 \\ 1 & \text{if} \quad x_j - x_k < 0 \end{cases} \qquad (2\text{-}4)$$

The random sequence $S_i (i = 1, 2, \cdots, n)$ approximates the normal distribution.

$$\mathrm{Var}(S) = \frac{1}{18}\left[n(n-1)(2n+5) - \sum_t t(t-1)(2t+5) \right] \qquad (2\text{-}5)$$

Using the following formula to calculate the statistical test value Z:

$$Z = \begin{cases} \dfrac{S-1}{\sqrt{\mathrm{Var}(s)}} & \text{if} \quad S > 0 \\ 0 & \text{if} \quad S = 0 \\ \dfrac{S+1}{\sqrt{\mathrm{Var}(s)}} & \text{if} \quad S < 0 \end{cases} \tag{2-6}$$

Where, Z is the statistic of a normal distribution and $\mathrm{Var}(S)$ is the standard deviation of S_i. $Z_a/2$ is obtained from the standard normal distribution function.

2.2.4.2 Sen's Estimator(S-E) nonparametric test

The S-E nonparametric test is a nonparametric test method used to estimate the actual variation of a variable(Sen, 1968). The S-E method can be used to assume that the variation of the variable conforms to the linear change. The actual variation of a linear variable in a time series can be estimated. This simple nonparametric test method is developed by Sen(1968).

It is estimated that the Q_i of N for the data amplitude can be calculated by the following equation:

$$Q_i = \frac{x_j - x_k}{j - k} \tag{2-7}$$

Where, x_j and x_k represent the values($j > k$) of j and k at a certain time, respectively.

The driving value of the n value of the S-E amplitude change Q_i is equal to the median thereof.

If n is an odd number, then the amplitude of S-E is calculated as:

$$Q_{\mathrm{med}} = \frac{Q_{(n+1)}}{2} \tag{2-8}$$

If n is an even number, then the amplitude of S-E is calculated as follows:

$$Q_{\mathrm{med}} = \frac{\left[\dfrac{Q_n}{2} + \dfrac{Q_{(n+2)}}{2}\right]}{2} \tag{2-9}$$

Where, Q_{med} is a nonparametric two-tailed test of the alpha confidence interval(Timo Salmi et al., 2002).

2.2.4.3 Sequential Version Mann-Kendall(S-M-K) nonparametric test method

Time series of trend tests often need to be combined with the change point test, with the S-M-K test method for mutation analysis, this test takes into account the time series($x_1, x_2, \cdots, x_n$) in all the relative value of the conditions. Assume that there is no tendency for zero hypothesis H_0 (observed x_i is randomly arranged over time) and another hypothesis H_1 is relative(there is a monotonically increasing or declining trend) (Feidas et al., 2004). x_j ($j = 1, 2, \cdots, n$) and x_k ($k = 1, 2, \cdots, j-1$), the number of $x_j > x_k$ is calculated and denoted by n_j in each set of comparisons.

The test statistic t_j is given by the formula:

$$t_j = \sum_{1}^{j} n_j \tag{2-10}$$

Test statistic mean and variance:

$$E(t) = \frac{n(n-1)}{4} \quad (2\text{-}11)$$

$$\mathrm{Var}(t_j) = \frac{[j(j-1)(2j+5)]}{72} \quad (2\text{-}12)$$

The sequence values of the statistics $u(t)$ are calculated by the following formula:

$$u(t) = \frac{t_j - E(t)}{\sqrt{\mathrm{Var}(t_j)}} \quad (2\text{-}13)$$

Where, $u(t)$ is a standard variable with zero mean and unit standard deviation. If $UF_k = u(t)$, then $UB_k = -u(t)$, if UF_k and UB_k two curves appear in the intersection, and the intersection between the critical lines, the intersection of the corresponding moment is the mutation point of the start time (Feidas et al., 2004).

2.2.4.4 Hurst index

It is important for managers and decision makers to understand the future trends of hydrological phenomena in the basin. The Hurst exponent has a strong ability to predict the future trend of time series. Therefore, this paper uses the Hurst index to predict the future trend of climate change in the study area. The method of estimating Hurst exponent includes absolute value method, aggregation variance method, R/S analysis method, periodic graph method, Whittle method, residual variance method and wavelet analysis method. The most widely used is the R/S analysis. R/S analysis is a nonparametric analysis without having to assume that the potential distribution is a Gaussian distribution, which makes little assumptions about the object of investigation and has good continuity (Li et al., 2008; Sakalauskiene et al., 2003; Zhao et al., 2011).

The principle of the R/S analysis is as follows: Consider a time series $\{x(j)\}, t = 1, 2, \cdots$ For any positive integer $j \geqslant 1$, define the very poor R sequence:

$$R(j) = \underset{1 \leqslant t \leqslant j}{\mathrm{Max}} X(t,j) - \underset{1 \leqslant t \leqslant j}{\mathrm{Min}} X(t,j) \quad (2\text{-}14)$$

Standard deviation S sequence:

$$S(j) = \left\{\frac{1}{j}\sum_{t=1}^{j}[x(t) - x_j]^2\right\}^{1/2} \quad (2\text{-}15)$$

If the standard deviation of the observed value is divided by the very poor (ie. the heavy standard deviation) to establish a dimensionless ratio, if the following relationship is. sati s fied:

$$\frac{R(j)}{S(j)} = (aj)^H \quad (2\text{-}16)$$

Where, α is a constant, then the time series exists Hurst phenomenon; H is Hurst index (Sakalauskiene, 2003; Xu Zongxue et al., 2007).

The value of H is in the range 0-1. When $H = 0.5$, as mentioned above, that is, the climate elements are completely independent, climate change is random. When $0.5 < H < 1$, it indicates that the time series has long-term correlation characteristics and the process is persistent. The overall trend of increasing the climatic elements in the past indicates that the overall trend in the future will increase or vice versa. And the H value is closer to 1, the stronger the persistence. When $0 < H < 0.5$, it indicates that the time series has long-term correlation. But the overall trend of the future is opposite to the past, that is, the overall increase in the past indicates that the overall reduction in the future, and vice versa, this phenomenon is anti-sustainability. The closer the H value is to 0, the stronger the persistence (Sakalauskiene, 2003; Xu et al., 2010)。

2.2.5 The time range and spatial scale of this study

According to the research objectives and data acquisition, the time scale of this study is as follows: Under the influence of natural conditions and human activities, the temporal and spatial distribution pattern of blue-green water in the Heihe River Basin (ie. Chapters 3 and 4) in 2005; the time range for the study of the changes in blue-green water in the Heihe River Basin (Chapter 5) and the spatial and temporal differences in blue-green water in the Heihe River Basin in the typical year (Chapter 6) is from 1960 to 2010.

This study uses the new boundary of the Heihe River Basin, and the spatial scale of this study is part of the Heihe River Basin in China. The division of the middle and lower reaches of this study is mainly based on the definition of the upper, middle and lower reaches of the Heihe River Basin by the Ministry of Water Resources (2002) of the People's Republic of China. In the present study, the sub-watershed above the Yingxiao Gorge is upstream; the sub-watershed between the Hengxia and the Yingxiao Gorge is the middle; the sub-basin below the Hengxia Gorge is downstream (see Figure 2-1).

2.2.6 Calculation method of blue and green water

In this study, blue-green water is based on the water balance formula (Arnold et al., 1998) and is calculated from the SWAT model output. In the SWAT model, the green water flow is the actual evapotranspiration (ET) for each hydrological response unit; the blue water flow is the sum of the surface runoff ($SURQ$) of the hydrological response unit, the lateral flow ($LATQ$), the underground runoff (GWQ).

The green water coefficient is calculated as follows:

$$GWC = \frac{g}{b + g} \tag{2-17}$$

Where, GWC is the the green water coefficient; g is a certain period (year, month, day) green water; b is a certain period (year, month, day) blue water.

The blue-green water depth of each sub-basin is calculated by the following formula:

$$S_{g/b} = \frac{\sum_{i=1}^{n} (w_i m_i)}{\sum_{i=1}^{n} m_i} \tag{2-18}$$

Where, $S_{g/b}$ is the total depth of water in a sub-basin (blue water, green water, blue water and green water); w_i is the blue-green water of a hydrological response unit; m_i is the area of a hydrological response unit; N is the number of hydrological response units in the sub-basin.

The following formula is used to calculate the relative change rate of blue-green water (RCR) in different periods.

$$RCR = \frac{(V_i - V_0)}{V_0} \times 100\% \tag{2-19}$$

Where, V represents the variable, in this study represents the blue water, green water, blue water and green water coefficient.

I represents the value of the variable in the i-th period, and 0 represents the value of the variable in the original period.

2.2.7 Water footprint evaluation method

The calculation method of WF_{green}, WF_{blue} and WF_{grey} in this study is based on the standard method provided by (Hoekstra et al., 2011).

2.2.7.1 Evaluation of blue water footprint and green water footprint

The blue-green water footprint of the crop is the sum of the blue-green water footprints of the crops in the watershed. The blue-green water footprints of each crop is obtained by the corresponding virtual water content (VWC) and the product of the respective production. The virtual water content refers to the amount of water (m^3) required to maintain crop growth (t) during crop growth, and the blue-green component in the virtual water is effective precipitation (ER, m^3/ha) or irrigation (IR, m^3/ha) and crop yield (Y, t/a). The virtual water content of the crop is the sum of the virtual water content (VWC_{green}) and the blue water virtual water content (VWC_{blue}).

$$VWC_{green} = \frac{ER}{Y} \tag{2-20}$$

$$VWC_{blue} = \frac{IR}{Y} \tag{2-21}$$

$$VWC = VWC_{green} + VWC_{blue} \tag{2-22}$$

The ER and IR of the crop were simulated by the CROPWAT model (FAO, 2010a; Allen et al., 1998), and the rain and irrigation were considered during the simulation. It is recommended to use the Irrigation Regime Act to calculate ER and IR (Hoekstra et al., 2011), because the method includes a daily equilibrium of soil moisture and a more accurate simulation of crop growth conditions. The study did not take into account the blue-green water contained in the crop itself, since the proportion of these components was very small (in general, these water contained only 0.1% of the moisture and only 1%) (Hoekstra et al., 2011).

The blue water footprint and green water footprint of livestock products are the sum of the blue water and green water footprints of all kinds of livestock products in the river basin. The blue water and green water trails of various livestock products are obtained by the corresponding product of virtual water content of meat and their respective products. Meat virtual water content refers to the amount of water (m^3) required for the production unit of meat (t).

Meat virtual water content consists of three parts: food water, livestock drinking water and product processing water (Mekonnen and Hoekstra, 2012). Livestock food mainly includes grass, feed and corn. Only maize cultivation in the Heihe River Basin requires both precipitation and irrigation, while the other two crops are mainly from precipitation (Zhang, 2003). The blue water and green water components of maize were simulated by CROPWAT model. Livestock drinking water and product processing water are from blue water. It is assumed that all food from livestock is derived from the Heihe River Basin. The food water consumption of a certain type of livestock (FWR, m^3/kg) is obtained by multiplying the food conversion factor of a crop (FCE_f, ratio of dry weight to output) by the virtual water content of the crop (VWC_f, m^3/kg).

$$FWR = \sum_{f=1}^{N_f} FCE_f \cdot VWC_f \tag{2-23}$$

$$VWC = FWR + DWR + PWR \tag{2-24}$$

Where, VWC is the amount of water used for livestock, m^3/t; DWR is the amount of the water consumption, m^3/t; PWR is the amount of the product processing water, m^3/t.

The food conversion coefficient, water consumption and product processing water required for food water use are from Zhang (2003).

In order to calculate the monthly water footprint of livestock products, it is assumed that the monthly drinking water is equal and the amount of water used for product processing is the same. Monthly food water and its blue water and green water components are mainly based on the CROPWAT model to calculate the monthly use of feeding crops.

The blue water footprint of the industrial sector and the living sector is obtained by multiplying the water consumption of each sector by the respective water consumption rate.

2.2.7.2 Evaluation of gray water footprint

The gray water footprint is measured by diluting the contaminants to the amount of water required to meet the environmental water quality standards. The formula is as follows:

$$WF_{grey} = \frac{L}{C_{max} - C_{nat}} \tag{2-25}$$

Where, WF_{grey} is the gray water footprint, m^3/yr; L is the pollutant discharge load (kg/yr); C_{max} is the highest concentration of pollutants (kg/m^3) in the case of environmental water quality standards; C_{nat} is the water initial concentration (kg/m^3), refers to the natural conditions of a pollutant concentration.

The agricultural sector is the main source of pollution for non-point source pollution. Non-point source pollution refers to dissolved or solid contaminants from non-specific sites, in the precipitation (or snow) scouring, through the runoff process into the receiving water (including rivers, lakes, reservoirs and gulf), and cause water eutrophication or other forms of contamination (Novotny et al., 1993). Such as fertilizers and pesticides applied in farmland, surface sediments accumulated in urban streets, etc. (Hao Fanghua et al., 2006). Non-point source pollution is more complex than point source pollution. The simplest calculation is to assume that a portion of the non-point source pollutant eventually reaches surface water or groundwater and the ratio of the substance entering the body to the substance(eg. , nitrogen element) is a fixed value(Hoekstra et al. , 2011) Ie leaching rate. The determination of the leaching rate is often dependent on the existing literature. The gray water footprint of agricultural non-point source pollution is calculated as follows:

$$WF_{grey} = \frac{L}{C_{max} - C_{nat}} = \frac{\alpha \cdot Appl}{C_{max} - C_{nat}} \tag{2-26}$$

In this formula, the variable *Appl* represents the amount of chemical applied, and α represents the proportion of the amount of contaminants caused by a substance entering the water body. For agriculture, nitrogen (N) or phosphorus (P) is usually used as a measure of the gray water footprint(Cheng et al. , 2010). In this case, α is the leaching rate of N or P.

The above model is simple and easy to use, can meet the purpose of preliminary rough calculation. Under the condition that the non-point source pollution data is not sufficient, this method can be used to estimate the gray water footprint of non-point source pollution.

Industrial and living sectors produce a lot of point source pollution during water use. Point source contamination refers to contamination of contaminants into water by identifiable sites (such as drainage lines, sewers, drains, etc. in factories and waste water treatment plants) and cause eutrophication or other forms of contamination of recipient water(Hill et al. , 1997). The gray water footprint of the industrial and living sectors is calculated using the general formula (2-6). Sewage usually contains a variety of forms of pollutants, gray water footprint determined by the most critical pollutants. The so-called most critical pollutants are the largest pollutants that cause the gray water footprint. For the industrial and living sectors, chemical oxygen demand(COD), ammonia(NH_3—H) and so on is the largest pollutant in the discharge of sewage, so often use COD, ammonia and other indicators as an indicator of industrial and life depart ment evaluate the gray water footprint.

Based on the gray water footprints generated by agriculture, industry and life, this study accounts for gray water footprints in different areas. In the process of accounting, because the number and type of pollution indicators selected by each department are different, and the water body can dilute different pollutants at the same time, the index with the maximum gray water footprint is defined as the pollution index of the department, and the gray water footprint quantity is the gray water footprint of the department. The calculation of the regional gray water footprint is similar to that of the departmental gray water footprint calculation method. After determining the gray water footprint and pollution index of each department, the gray water footprint of the department with the same pollution index is added, and the larger gray water footprint is selected as the total gray water footprint in the area.

In the calculation of gray water footprint in the Heihe River Basin, the agricultural sector uses N and P as the pollution index, and the industrial and living sectors are in the industrial and living sectors. COD and NH_3—N were the pollution indexes. NH_3—N, COD, TN(total nitrogen) and TP (total phosphorus) were the main indexes in the three major river basins and provinces.

2.2.7.3 Evaluation method of water shortage based on water footprint

The water shortage index (I) is an indicator of the degree of shortage of water resources, which is a description of the specific water and water quality indicators. It is defined as the sum of the water quality indicators (I_{blue}) and the water quality indicators (I_{grey}).

$$I = I_{blue} + I_{grey} \tag{2-27}$$

I is defined as the ratio of water consumption (surface water and groundwater use) of specific areas in a specific period to the amount of freshwater resources. Similar to the stress coefficient method proposed by Alcamo et al. (2000). I_{blue} threshold is chosen to be 0.4, ie., when an area of I_{blue} is above 0.4, it indicates that the water shortage in this area is already very severe (Vörösmarty et al., 2000; Alcamo et al., 2003; Falkenmark and Rockström, 2004; Oki and Kanae, 2006). In general, since 80% of natural runoff is required to maintain the health of the environmental flow (Hoekstra et al., 2011), in fact, when the I_{blue} value is higher than 0.2, it indicates that the local water quality problem has been faced.

$$I_{blue} = W/Q \tag{2-28}$$

I_{grey} is a measure of water quality and water shortage, defined as the ratio of WF_{grey} (m^3/yr) to freshwater resources (Q, m^3/yr) in a given area. If I_{grey} is less than 1, it indicates that the actual use of fresh water can dilute existing pollution based on local water quality standards; conversely, the actual available fresh water can not completely dilute the local pollution, so the threshold for I_{grey} is defined as 1.

$$I_{grey} = WF_{grey}/Q \tag{2-29}$$

2.2.7.4 Evaluation method of water shortage in Heihe River Basin

The study of the Heihe River Basin mainly analyzes the shortage of local water resources by the method shown in Section 2.3, and also makes a sustainable evaluation of the blue water footprint in the Heihe River Basin.

Blue water sustainability indicators were obtained by comparing the blue water footprint and blue water availability within the watershed. When the blue water footprint exceeds the available amount of blue water, attention is needed to the sustainability of the area (Hoekstra and Mekonnen, 2012b). When the sustainability index is less than 100%, it indicates that blue water is sustainable. When the sustainability index is 100%, it indicates that blue water is unsustainable and the higher the sustainability index is, the stronger the blue water is unsustainable. The amount of blue water available (WA_{blue}) is calculated as follows:

$$WA_{blue} = BWR - EFR \tag{2-30}$$

Where, BWR is the amount of blue water resources in nature, ie. natural runoff, which is equal to the sum of surface water and groundwater; EFR is the ambient flow.

The Heihe River Basin across China's three provinces, 15 cities and counties. Due to the lack of Mongolian data, coupled with the Heihe River Basin in the country within the natural situation is mainly desert, human activities carried out less, ignore the region and the Heihe River Basin will not have a significant impact on the overall water footprint evaluation, so this study only considers the Heihe River Basin in China. Because it is not possible to obtain data from watersheds directly, data from the basin scale is obtained by combining the statistical data in the administrative area with the spatial data.

When calculating the blue water footprint and green water footprint in the Heihe River Basin, the crop and livestock product production data in the river basin are needed. The statistical data only provide the crop harvest area and yield of the 15 cities and counties in 2004-2006, and there is no data of the river basin scale. Combined with these statistics, and the crop distribution with a precision of 5 arcs obtained from the University of Frankfurt's MIRCA2000 database (Portmann et al., 2010), the percentage of acreage of a crop in the administrative area covered by the Heihe River Basin (including rainfed area and irrigated area) to calculate the total acreage of a crop in the Heihe River Basin. Thus, the crop area data of the various crop types at the basin scale can be obtained. The crop yield data for the crop scale are obtained in the same way. Table 2-1 shows the harvested area and yield of various crops in the Heihe River Basin.

A total of 12 typical crop types in the Heihe River Basin were selected and evaluated. Each crop has its representative crop (see Table 2-1). The crop types include cereal crops (wheat, corn and other cereal crops), soybeans, potato crops (potatoes), oil crops (rapeseed), carbohydrates (beets), cotton, fruits (apples and other fruits), vegetables (Tomatoes) and other crops. By calculation, the first 11 crops accounted for 86% of Heihe total crop yield and 14% of other crops.

Production of livestock products (meat) is calculated primarily by the number of livestock of a class multiplied by the average meat production per head of livestock. Beef, lamb, pork and poultry meat are four types of major livestock products in the Heihe River Basin. The study only calculates and evaluates four types of livestock products. The density data for each type of livestock comes from the FAO Food and Health Distribution Database (FAO, 2011). The database provides spatial information on the density of livestock in 2005 with a spatial accuracy of 3 arc points. The total number of livestock in the Heihe River Basin is summed up by the number of such livestock in all rasters in the basin in the spatial distribution map.

In the calculation of the blue-green water footprint of the Heihe River Basin crop, the relevant data needed to simulate the virtual water content of the crop using the CROPWAT model (FAO, 2010a; Allen et al., 1998). The CROPWAT model requires climate, crop and soil-related data for crop evaporation and irrigation simulations. Climate data include temperature, precipitation, humidity, light, radiation and wind speed. The climate data are from the New LocClim database (FAO, 2005), which provides 30-year average monthly weather data (1961-1990). This study is based on data from three meteorological stations in the Heihe River Basin (coordinates 98.3E, 38.4N; 99.4E, 39.1N; 100.6E, 38.4N, respectively). Please refer to Chapter 4 for site location. Crop parameters such as crop coefficient, root depth, crop growth stage, planting and harvest date, etc. from Allen et al. (1998) and Chapagain and Hoekstra (2004). Soil parameters include soil available water content, maximum infiltration rate of different soil types, maximum root depth, and initial soil water content. The soil water content in the Heihe River Basin is derived from FAO's Global Soil Property Map (2010b), with maximum infiltration rates using sand and loam data for major soil types in Heihe (Qi and Cai, 2007). Due to the lack of information on the maximum root depth and initial soil moisture content of the crop growing season in the basin, the default value in the CROPWAT model is selected for the parameter (FAO, 2010a).

Table 2-1 Annual harvested area and crop production within the HRB (2004-2006)

Crop type	The represents the crop	Harvest area (×1,000 ha)	yield (×1,000 t)
wheat	wheat	53	322
Corn	Corn	30	239
Other cereal crops	Barley	50	352
Soybean	Soybeans	3	21
Potatoes	Potatoes	11	87
Oil crop	rapeseed	18	47
Sugar Beet	Beets	8	190
Cotton	Cotton	21	46
Apples	Apples	5	27
Other fruit	pears	45	229
Vegetables	Tomatoes	27	740
Other crops			366

The total water consumption in the Heihe River Basin is about 34.33×10^8 m^3/yr (Xiao Honglang et al., 2006), where water consumption is 44.20×10^6 m^3/yr, industrial water is 95.20×10^6 m^3/yr (Chen et al., 2005). The department's water consumption rate was 67% and the industrial sector had a water consumption rate of 36% (GSMWR, 2006). The annual runoff and monthly runoff data for the Heihe River Basin for the sustainable evaluation of blue water in the Heihe River Basin are from Zang et al. (2012). They use the soil and water evaluation model—SWAT model (Arnold et al., 1994) The surface water and groundwater were simulated. The environmental flow is part of the natural runoff, using the global average of 80% of the proposed global level of Hoekstra et al. (2011, 2012b) as the ratio of ambient and natural runoff.

In the calculation of the gray water footprint of the relevant departments of the Heihe River Basin, the sewage data of each department in the river basin (ie., the application amount of agricultural nitrogen and phosphate fertilizer, COD and NH_3—N in the industrial and living sectors) and the pollution index maximum allowable concentration and natural concentration. The data required by the various departments in the river basin can not be directly obtained. The required data are calculated from the statistical data and spatial data of the relevant provinces and cities from 2004 to 2006. According to the statistical yearbook of Gansu, Qinghai and Inner Mongolia, the acreage of crops in 15 cities and counties in the Heihe River Basin can be obtained. The spatial distribution of crops in the Heihe River Basin is obtained. After the province's nitrogen and phosphate fertilizer application in accordance with the proportion of acreage allocated to the Heihe River Basin, the Heihe River Basin nitrogen fertilizer and phosphate application. As the counties and cities in the Heihe River Basin are mainly mountainous areas, the counties and cities involved in the Inner Mongolia Autonomous Region are mainly desert areas, the population is relatively small, the industrial and domestic sewage is not obvious. The industrial and domestic sewage in the river basins are mainly concentrated in Gansu, therefore, according to the proportion of the population of the Heihe River Basin and Gansu Province, the COD and NH_3—N emission from the industrial and living sectors of Gansu Province were allocated to the Heihe River Basin, and the sewage and industrial data of the Heihe River Basin were obtained. The maximum allowable concentrations of COD and NH_3—N are 20 mg/L and 1 mg/L, respectively, and the N element and the P element are 10 mg/L, respectively, according to the Class Ⅲ water quality standard for the Standard Water Resources Quality Standard-1 is 2 mg/L, and the natural concentration of the four pollution targets is assumed to be 0 (Hoekstra et al., 2011). Due to the lack of N and black elements in the Heihe River Basin, select the global average of 10% and 2% as the local leaching rate (Hoekstra et al., 2011).

Chapter 3 Study on Spatial and Temporal Dynamic Distribution Pattern of Blue-Green Water in The Heihe River Basin under Natural Conditions

3.1 Research background

Ensuring sufficient water supply is essential for the survival and sustenance of humans and ecosystems(Oki and Kanae,2006). However, with population growth and socioeconomic development, more and more water is used to solely meet human requirements. This often leads to decreasing water availability for ecosystem use with implications for ecosystem health. In the long term, insufficient water availability for essential ecosystem functions and services can lead to ecosystem degradation with consequent impacts on overall water scarcity and human wellbeing (Falkenmark,2003). In particular in arid and semi-arid regions, water use competition is intense between human and ecosystems; hence, a comprehensive assessment of water resources in a spatially and temporarily explicitly way is a key to deepening the understanding of the renewable water endowments as well as to enhancing water management towards sustainable, efficient and equitable use of limited water resources.

Traditionally, water resources assessment and management have put emphasis on blue water, ignoring green water(Falkenmark,1995a; Cheng and Zhao,2006). Conceptually, water can be divided into green water and blue water(Falkenmark,1995a). Blue water is the water in rivers, lakes, wetland and shallow aquifers, while green water is precipitation water stored in unsaturated soil, and later used for evapotranspiration. Although green water is often ignored, it plays an essential role in crop production and other ecosystem services. J. Liu et al. (2009) estimated that green water accounts for more than 80% of consumptive water use for global crop production. Rost et al. (2008) estimated that green water consumption in global cropland to be from 85% in 1971 to 92% in 2000 of total crop water consumption. Green water dominates water uses in tropical arid regions, where rainfed agriculture accounts for more than 95% of total cropland area (Rockstrtöm, 1999). Water use in grassland and forest ecosystems is dominantly "green".

Since the concept of green and blue water was introduced(Falkenmark, 1995a, 1995b), green/blue water research has become more and more diversified, especially after Falkenmark and Rockström (2006) conceptualized a wider green-blue flows approach for water-resource planning and management. Many novel research methods have appeared as well. For instance, Rost et al. (2008) and Gerten et al. (2005) use the LPJmL model to assess global green water consumption over a time period of nearly 30 years, while J. Liu et al. (2009) used the GEPIC model to calculate the global green/blue water consumption of cropland. Schuol et al. (2008) and Faramarzi et al. (2009) used the SWAT model to simulate green/blue water resources of Africa and Iran, respectively. The green/blue water concept has offered a new methodology and fresh ideas for water resources management in many regions, in particular in arid and semi-arid regions where water scarcity is serious due to water-thirsty socioeconomic development and population growth. Novel measures and concepts can aid in underpinning more sustainable and equitable water resources management (Jansson et al., 1999).

Importantly, and especially for data poor regions, hydrological flows under conditions unaffected by human activities are often poorly characterized. Many studies tend to pay attention to the influence of human activities, but generally fail to characterize the state of the ecosystem under natural conditions. Modelling tools can aid in representing natural conditions and can be used as reference for follow-up studies and inform researchers and policy makers about the original state of a river basin as an input into decision-making.

The Heihe River is the second largest inland river in China, located in Northwest China, it originates in the Qilian Mountains and discharges into the Juyanhai Lake. The Heihe River Basin is a typical arid and semi-arid region, suffering from a serious water crisis (Cheng et al., 2006). Water use in mid-stream regions has increased sharply in the Heihe River Basin related to socio-economic development (Ma et al., 2011). As a consequence, the Heihe River Basin has been confronted with serious ecosystem degradation including the complete dry-up of the downstream west and east Juyanhai Lakes (Cheng, 2002). Other related environmental crises in the area include the southward expansion of the Badain Jaran desert and an increased occurrence of sand-storms (Li, 2009). So far, the main measures of water resources manage-ment in the Heihe River Basin include water transfer, irrigation and a hydropower project (Xiao et al., 2011). Most of the water management has paid attention to the liquid blue water, while stored green water has been ignored.

Furthermore, to formulate water management to be in line with environmental capacity a good understanding of the spatiotemporal patterns of the natural hydrological flows in the basin is needed. Unfortunately, this type of information and subsequent interpretation and analysis are currently lacking for the Heihe River Basin. Therefore, the aim of our current research is to establish a benchmark for the natural flows of water in the basin and to quantify the spatial and temporal dynamics of green water and blue water in the entire Heihe River Basin under natural conditions. This information can provide a reference for subsequent studies and it can also be used to inform policy makers about the original state of water resources in the basin. Importantly, we focus on the complete river basin opposed to other studies that have looked at certain segments of the basin which are not sufficient to achieve integrated green water and blue water management. Specific objectives were: ①to calibrate and validate the SWAT model at two hydrological stations accounting for 85% of the total discharge in the Heihe River

Basin but that are not much affected by human intervention;②to quantify the spatial and temporal dynamics of green water and blue water under natural conditions in the entire Heihe River Basin and discuss implications for further research.

3.2 Methodology

The Heihe River Basin lies between longitudes 97°05′-102°00′ E and latitudes 7°45′-42°40′N. With a total basin area of 0.234 million km^2, this river basin is mainly located in Northwest China, but it also has a part in Mongolia (see Figure 3-1). There are two often-used river basin boundaries. The old one has an area of 0.116 million km^2. Such a boundary was created based on administrative boundaries (mainly the boundaries of different counties), but it lacked a practical hydrological sense. Realizing this, the Heihe Data Research Group has worked on a more accurate and complete new river basin boundary by integrating hydrological simulation with measured river system data (http://www.westgis.ac.cn/datacenter.asp). The output is the new river basin boundary with an area of 0.234 million km^2. Such a boundary not only reflects a more accurate division with an explicit hydrological meaning, but also reflects a watershed boundary under natural conditions. The average altitude of the basin is over 1 200 m. With a total length of 821 km, the Heihe River is divided into three sections: upstream, midstream and downstream. The upstream runs from the Qilian Mountain to the Yingluo Canyon with a length of 303 km, the mid-stream runs from the Yingluo Canyon to Zhengyi Canyon, while the downstream goes from the Zhengyi Canyon and terminates into the Bada in Juran desert. The annual temperature is 2-3 ℃ upstream, 6-8 ℃ midstream, and 8-10 ℃ downstream. The average annual precipitation is between 200 mm to 500 mm in the upstream, less than 200 mm in the midstream, and less than 50 mm in the downstream area. Potential evaporation ranges from 1,000 mm/yr upstream to 4,000 mm/yr downstream (Liu et al., 2008). The precipitation in the Heihe River Basin occurs mainly in summer, spring and autumn are dry and some melting water is generated in spring, there is much snow in winter. The climate of the Heihe River Basin is very dry, especially downstream where the drought index defined by the ratio of potential evapotranspiration and precipitation equals 47.5 (Li, 2009). There are 24 tributary channels with a total annual runoff larger than 10 million m^3, with more than 0.375 billion m^3 coming from the Qilian Mountains. The multiyear annual average precipitation is 122.6 mm, and most of this falls between May and August, accounting for over 70% of the total annual precipitation (Li, 2009). The melting water of the Heihe River Basin amounts to 0.1 billion m^3, or 4% of the total discharge (Li, 2009). Since the 1980s, agricultural water use has increased sharply midstream. From the 1980s to the 1990s, the annual discharge through the Zhengyi Canyon decreased from 0.942 billion m^3 to 0.691 billion m^3 (Xiao et al., 2011). River runoff provides about 65% of the irrigation water in midstream region while groundwater provides over 90% of the irrigation water in the downstream region (Xiao et al., 2011). The main land cover types are desert, mountains and oasis, which cover 57.15%, 33.16% and 8.19% of the total basin area, respectively(Cheng et al. ,2006).

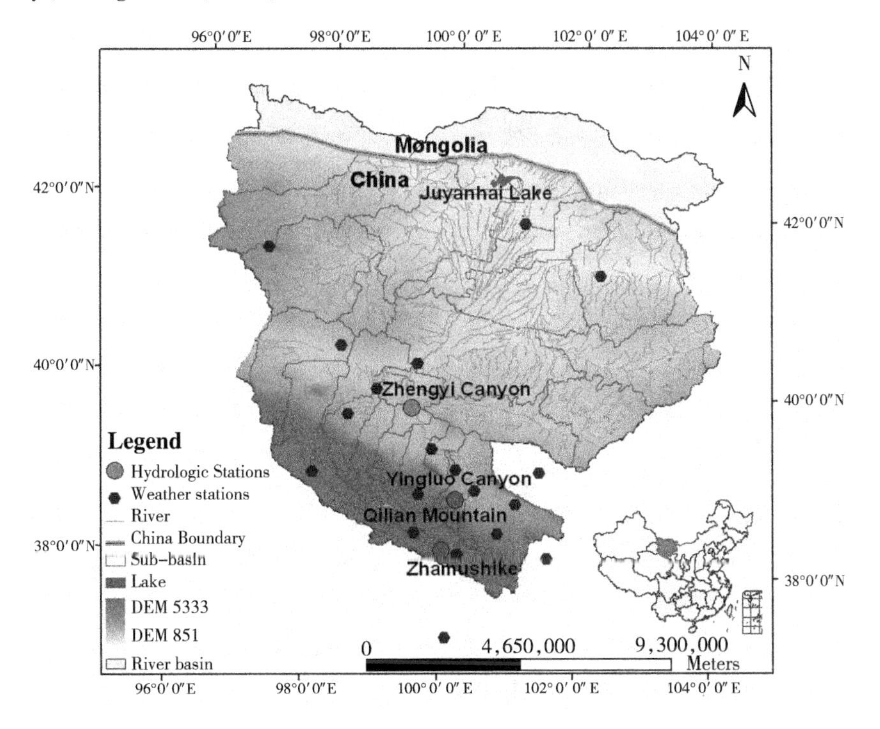

Figure 3-1 The Heihe River Basin with DEM, rivers, hydrological, lakes and weather stations indicated. The location of the Heihe River Basin in China is shown in the inset

The Heihe River Basin has complex ecosystems ranging from mountains in the South, oases in the middle and deserts in the North(Cheng et al. ,2006). These ecosystems are linked from upstream to downstream by the water cycle. In recent years, with socio-economic and population development, the water flow through the Heihe River Basin has been diminishing year by year. For example, Zhangye, the biggest city of the Heihe River Basin located midstream, has witnessed a population increase of 14,000 persons per year, with the population amounting to 1.27 million in 2000. Irrigated agricultural area has increased by 2.87 thousands ha/yr, with the total irrigated area reaching 216 thousands ha in 2007(Liu et al. ,2008). Therefore, a detailed and integrated simulation study of the water resources of the complete river basin is critical and urgent for better water management.

3.3 Spatial and temporal distribution of blue-green water under natural conditions

3.3.1 The rate of the model and the verification results

Green-blue water can refer to both volume and flow. Here the flow concept is taken. Green water flow refers to actual evapotranspiration, while blue water flow is the sum of surface runoff, lateral flow, and return flow from shallow aquifers. The green water coefficient (GWC) is defined as the ratio of green water flow to the total green water and blue water flows, and it is calculated by the equation below (J. Liu et al., 2009).

$$GWC = \frac{g}{(b + g)} \tag{3-1}$$

Where, b and g are blue water and green water flows, respectively, in mm/yr.

The relative change rate (RCR) is used to indicate the change of green/blue water flows in different periods.

$$RCR = \frac{v_j - v_0}{v_0} \times 100\% \tag{3-2}$$

Where, v refers to the variables such as green water flow or blue water flow; i indicate the latter period and 0 indicates the initial period.

3.3.2 The total blue-green water changes in the Heihe River Basin

We use the Soil and Water Assessment Tool (SWAT) to simulate green water and blue water flows. The SWAT model is a semi-physically based, semi-distributed, basin-scale model (Neitsch et al., 2004), which has been used widely in many countries around the world (Schuol et al., 2008; Gerten et al., 2005; Faramazi et al., 2009). There are two main reasons for selecting the SWAT model. Firstly, it has already been successfully applied for water quantities (Faramazi et al., 2009; Schuol et al., 2008) and quality (Gassman et al., 2007) assessments for a wide range of scales and environmental conditions, including green-blue water assessments; secondly, the SWAT model has been used to simulate the hydrological processes in a small upstream segment of the Heihe River Basin successfully (Huang and Zhang, 2004; Li et al., 2009). There are more than nine types of hydrological models that have been used in the Heihe River Basin for water resources research (Li, 2009). Nevertheless, all of these model simulations have focused on upstream river segments in the Qilian Mountains, which form only 14.7% of the total river basin area. The hydrological processes have never been studied for the entire river. An important reason is that past research on hydrological cycles is often focused on human water use, particularly blue water use, thus overlooking water use by ecosystems. The up-and middle segments are regions where blue water is generated and used, but the downstream segments and surrounding areas are dominated by natural ecosystems and a low population density. Hence, most of the studies have focused on simulating upstream seg-ments and not the entire basin or downstream watersheds. However, we argue that studying the hydrological processes for the entire basin is essential since water is not only required by human beings but also needed by natural ecosystems. In addition, a study covering the entire basin makes more sense from a hydrological point of view. An additional reason for the emphasis on upper river segments may also be the lack of available data for the downstream river segments. In our research, we use SWAT2005, which was running on Arcview 3.3 with a daily time step. In SWAT, the modelled area is divided into multiple sub-basins and hydrological response units (HRUs) by overlaying elevation, land cover, soil, and slope classes. The HRUs are characterized by combinations of dominant land-use, soil, and slope classes. This choice was essential for keeping the size of the model at a practical limit. For each of the sub-basins, water balance was simulated for four storage volumes: snow, soil profile, shallow aquifer, and deep aquifer. Potential evapotranspiration was computed using the Hargreaves method (Hargreaves et al., 1985). The calculation of evaporation requires the input of daily precipitation, and minimum and maximum temperature. Surface runoff was simulated using a modified SCS curve number (CN) method and snow and melting water cal-culated by the energy balance equation. Further technical model details are given by Neitsch et al. (2004). The preprocessing of the SWAT model input was performed within ESRI ArcGIS 9.3.

The Av-SWAT interface was used for the setup and parameterization of the model. The entire river basin was divided into 303 HRUs and 34 sub-basins on the basis of the digital elevation model (DEM). The geomorphology, stream parameterization, and overlay of soil and land cover were automatically done within the interface. We only present results for the river basin within the Chinese boundary due to the lack of data for Mongolia.

3.3.3 The blue-green water dynamics in the middle and lower reaches of the Heihe River Basin

The SWAT model mainly requires five types of data: DEM, land use data, soil data, climate data, and other management data. A large part of the data for the Heihe River Basin was obtained from the Heihe Data Research Group. The collection of the data was followed by an accuracy assessment and analysis of the quality and integrity of the data. The basic input maps included DEM at a resolution of 30 m (USGS/EROS, 2009) and land cover at a resolution of 1 km from the Heihe Data Research Group. There are 26 classes of land use in the Heihe River Basin including cropland, forest, grassland, lakes, wetland, among others. We have built the land use database using Chinese land cover type characters. The soil data was obtained from the Harmonized World Soil Database produced by the Food and Agriculture Organization of the United Nations (FAO), the International Institute for Applied Systems Analysis (IIASA), and the Institute of Soil Science Chinese Academy of Sciences (ISSCAS) (http://www.iiasa.ac.at). This dataset has a spatial resolution of 30 arc-second (about 1 km), and it includes 63 soil types for the Heihe River Basin with two soil layers (0-30 cm and 30-100 cm depth) for each type. The climate data for 19 weather stations were used for model simulation (see Figure 3-1). The daily climate input data (precipitation, minimum and maximum temperature) for the period of 1977-2004 were obtained from the Heihe Data Research Group and China Meteorological Data Sharing Services System (http://cdc.cma.gov.cn/index.jsp). River discharges for a time period from 1977-1987 and 1990-2004 were also provided by the Heihe Data Research Group. As a first step, we aim to simulate green and blue water flows without human intervention; hence, management data such as irrigation were not collected.

Model calibration and validation is a challenging and to a certain degree subjective step in a complex hydrological model. We aim for the model simulation to reflect natural conditions. Therefore, the SWAT model of the Heihe River Basin was calibrated and validated using monthly river discharges for two upstream stations where human activities are not intensive. These stations are the Zhamushike station and Yingluo canyon (see the locations in Figure 3-1) and have also been used for calibration and validation by Li (2009). Our reasons are two-fold: ① more than 85% of the annual discharge in the Heihe River flows through these two hydrological stations so the optimised parameters of the upstream area will be very important in representing the entire watershed; ② the stations are not much affected by human interference which is in line with our aim of analysing the green and blue water distribution under natural conditions. Furthermore, they have the most complete discharge data for 1977-1987 and 1990-2004.

The simulation period was from 1977 to 2004. The first two years were used as warm-up period to mitigate the effect of unknown initial conditions, which were subsequently excluded from the analysis. Hence, we divide the discharge data into two periods: a calibration (1979-1987) and a validation period (1990-2004).

Based on the built-in sensitivity analysis tool in SWAT (Neitsch et al., 2004), we have identified the 11 most sensitive parameters. In addition, based on previous studies, three other parameters (SMFMX, SMFMN and TIMP in Table 3-1) were identified as also being important for SWAT simulation in the Heihe River Basin (Li, 2009). These 14 parameters are listed in Table 3-1. Two indices, the Nash-Sutcliffe coefficient (Eq. (3-3)) and the coefficient of determination (Eq. (3-4)), are used to evaluate the goodness of the calibration and validation.

Table 3-1 The most sensitive parameters and their best parameter intervals and values

Aggregate Parameter*	Description	Best parameter interval	Best parameter value
r_CN2	Initial SCS CN Ⅱ value	0.47-0.59	0.51
v_ALPHA_BF	Base-flow alpha factor[days]	0.92-0.99	0.94
v_GW_DELAY	Groundwater delay[days]	462-473	467
v_GWQMN	Threshold water depth in the shallow aquifer for flow[mm]	0.72-0.85	0.77
v_GW_REVAP	Groundwater" revap" coefficient	0.094-0.11	0.098
v_ESCO	Soil evaporation compensation factor	0.78-0.80	0.79
v_CH_K2	Channel effective hydraulic conductivity[mm/h]	23-29	27
r_SOL_AWC(1)	Available water capacity[mm H_2O mm/soil]	0.11-0.18	0.14
r_SOL_K(1)	Maximum canopy storage[mm]	0.22-0.23	0.23
v_SFTMP	Snowfall temperature[℃]	-1.87-1.41	0.79
v_SURLAG	Surface runoffl ag time[days]	4.18-5.19	4.68
v_SMFMX	Melt factor for snow on 21 June[mm H_2O ℃/day]	5.85-6.27	6.02
v_SMFMN	Melt factor for snow on 21 December[mm H_2O ℃/day]	3.05-3.51	3.25
v_TIMP	Snow pack temperaturel ag factor	0.38-0.622	0.49

Note: * The aggregate parameters are constructed acoording to Yang's work (Yang et al, 2007, 2008). "v_" "r_" means an in crease, a replacement and a relative change to the initial parameter value respectively. The range of the aggregate parameter best distribution for is mainly b ased on SWAT-CUP calibration results.

$$E_{ns} = 1 - \frac{\sum_{i=1}^{n}(Q_{oi} - Q_{mi})^2}{\sum_{i=1}^{n}(Q_{oi} - \overline{Q}_o)^2} \tag{3-3}$$

$$R^2 = \frac{\left[\sum_{i=1}^{n}(Q_{oi} - \overline{Q}_o)(Q_{mi} - \overline{Q}_o\right]^2}{\sum_{i=1}^{n}(Q_{oi} - \overline{Q}_o)^2(Q_{mi} - \overline{Q}_o)^2} \tag{3-4}$$

Where, E_{ns} is the Nash-Sutcliffe coefficient; Q_{oi} is the observed data of runoff in i years; Q_{mi} is the simulation data of runoff in i years, and n is the length of the time series.

The closer E_{ns} and R^2 are to 1, the more accurate the model prediction, whereas an $E_{ns} > 0$ indicates that the model is a better predictor than the mean of the observed data. More information about the Nash-Sutcliffe coefficient and SWAT-CUP can be found in Nash and Sutcliffe (1970) and Abbaspour (2007), respectively.

The SUFI-2 method in the SWAT-CUP interface (Abbaspour et al., 2007) was used for parameter optimization. In this method all uncertainties (parameter, conceptual model, input, etc.) are mapped onto the parameter ranges, which are calibrated to bracket most of the measured data in the 95% prediction uncertainty (Abbaspour et al., 2007). The overall uncertainty analysis in the output is calculated by the 95% prediction uncertainty (95PPU) and we chose two different indices to compare measurement to simulation: the P-factor and the R-factor. The P-factor is the percentage of data bracketed by the 95PPU band. The maximum value for the P-factor is 100%, and ideally we would like to bracket all measured data, except the outliers, in the 95PPU band. The R-factor is the average width of the band divided by the standard deviation of the corresponding measured variable (Abbaspour, 2007; Faramarzi et al., 2009). The R-factors were calculated as the ratio between the average thickness of the 95PPU band and the standard deviation of the measured data. It represents the width of the uncertainty interval and should be as small as possible. R-factor indicates the strength of the calibration and should be close to or smaller than a practical value of 1 (Abbaspour, 2007).

3.3.4 Temporal and spatial distribution of green water coefficient in the Heihe River Basin

The calibration and validation performed with SWAT at the two hydrological stations was satisfactory, as indicated by high values of E_{ns} and R^2 (see Figure 3-2). The E_{ns}-values at both Zhamushike and Yingluo canyon are above 0.87, and the R^2-values are greater than 0.90. Our calibration and validation results seem better than those from Huang and Zhang (2004) and Li (2009). Meanwhile, the simulated and observed discharges have very similar variation trends (Figure 3-2), especially in the validation period of Yingluo Canyon.

The good agreement between the simulation results and observations indicates that the SWAT model set-up is suitable for the Heihe River Basin. The most sensitive parameters with

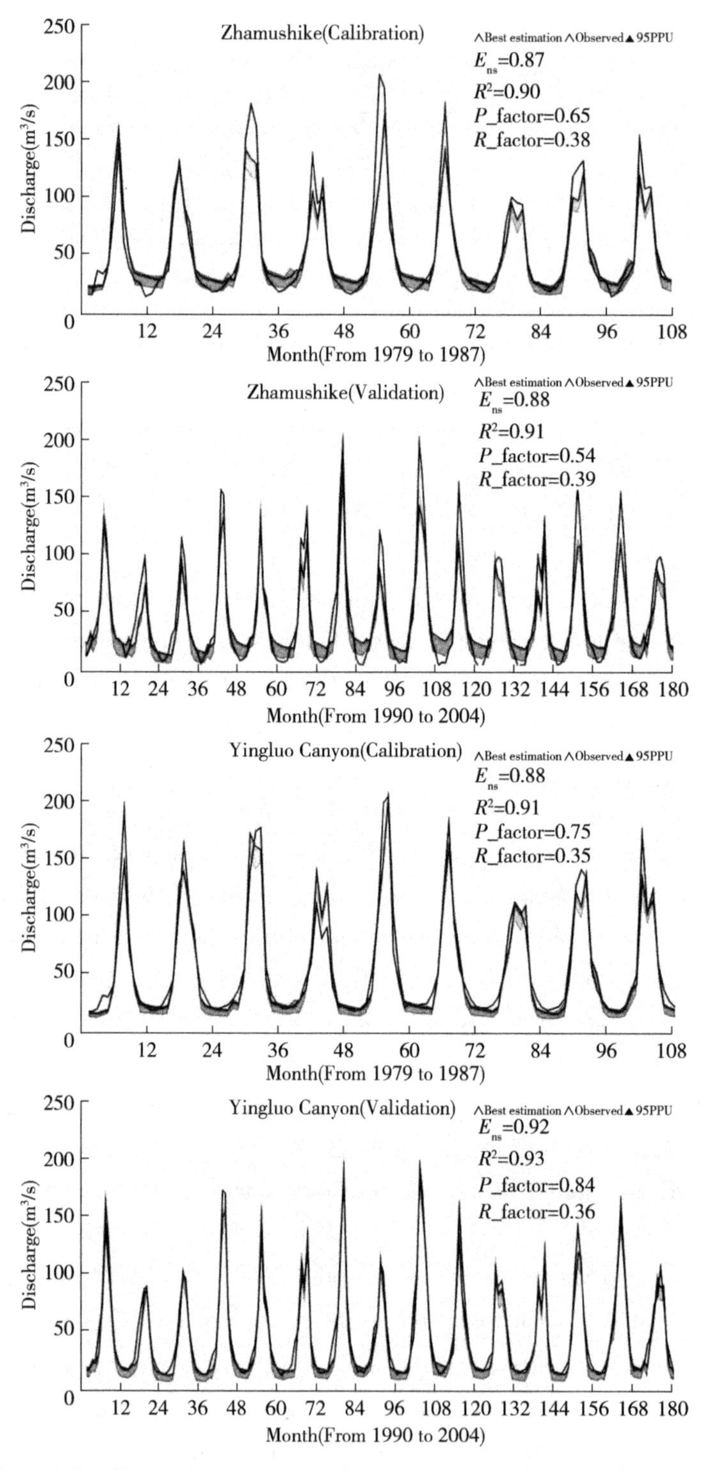

Figure 3-2 Comparisons between the observed and simulated (expressed as 95% prediction uncertainty band) discharge for the Zhamushike and Yingluo canyon hydrological stations in the Heihe River Basin

their best parameter intervals and best parameter values eventually used in this study are shown in Table 3-1.

Nevertheless, several challenges remain while optimizing the parameters, for instance, model calibration using river discharge alone does not provide confidence on the partitioning of water between soil storage, actual evapotranspiration and aquifer recharge. When additional datas (e. g. measured evapotranspiration) are available, a multi variable calibration is required to calculate water resources availability based on water yield and green water components.

The spatial and temporal distribution of total water flow (sum of green and blue flows) in the Heihe River Basin is showed in Figure 3-3. From the relative change rate, we found that there is a general decreasing trend in per unit area water flow (in mm/yr) from upstream to downstream sub-basins (see Figure 3-3). This is easy to understand because annual precipitation de-creases from upstream to downstream with snow and melting water upstream (Wang and Zhou, 2010).

The total water flow was 22.05-25.51 billion m^3 in the 2000s for the entire river basin. There are several blue colored regions that stand out with relative high total water flow in volume: those upstream generally have high precipitation and often a large volume of snow and melting water (Li, 2009), while those downstream are often resulting from large sub-basin areas. SWAT generates sub-basins based on DEM, land use and soil types. Because downstream regions have more homogeneous distribution of elevation, land use and soil types, the downstream sub-basin areas can be ten times larger than those upstream. From the 1980s to the 1990s, the total water flow has a general decreasing trend in upstream and midstream sub-basins, but the relative change rate has a general increasing trend in downstream sub-basins. However, for the relative change rate from the 1990s to 2000s, there are very different change patterns, with increasing trends in upstream and middle stream sub-basins but decreasing trends in downstream sub-basins (see Figure 3-3). In upstream and midstream sub-basins, precipitation and temperature had decreasing trends from the 1980s to 1990s, but increasing trends from 1990s to 2000s (Wang and Zhou, 2010), leading to different change patterns in total amount of precipitation water and snow and melting water. In downstream subbasins, sunshine durations increased from the 1980s to the 1990s but decreased from the 1990s to the 2000s (Y. Liu et al., 2009), which caused increasing and decreasing temperature in the two periods, respectively. The temperature variation caused evapotranspiration changes downstream (Cheng et al., 2007). Therefore, climate variability is a main reason for the variation of total water flow under natural conditions in the Heihe River Basin. From 1980s to 2000s, the total water flow of Heihe River Basin did not change much with a very slight increase by about 1.1%-1.4% (see Figure 3-4).

Both the green water and blue water flows per unit area in the Heihe River Basin decreased from upstream to downstream (see Figure 3-5). Generally, where blue water flows per unit area are high, green water flows also tend to be high (see Figure 3-5), in line with findings of previous research (Schuol et al., 2008). The spatial patterns of the green/blue water flows

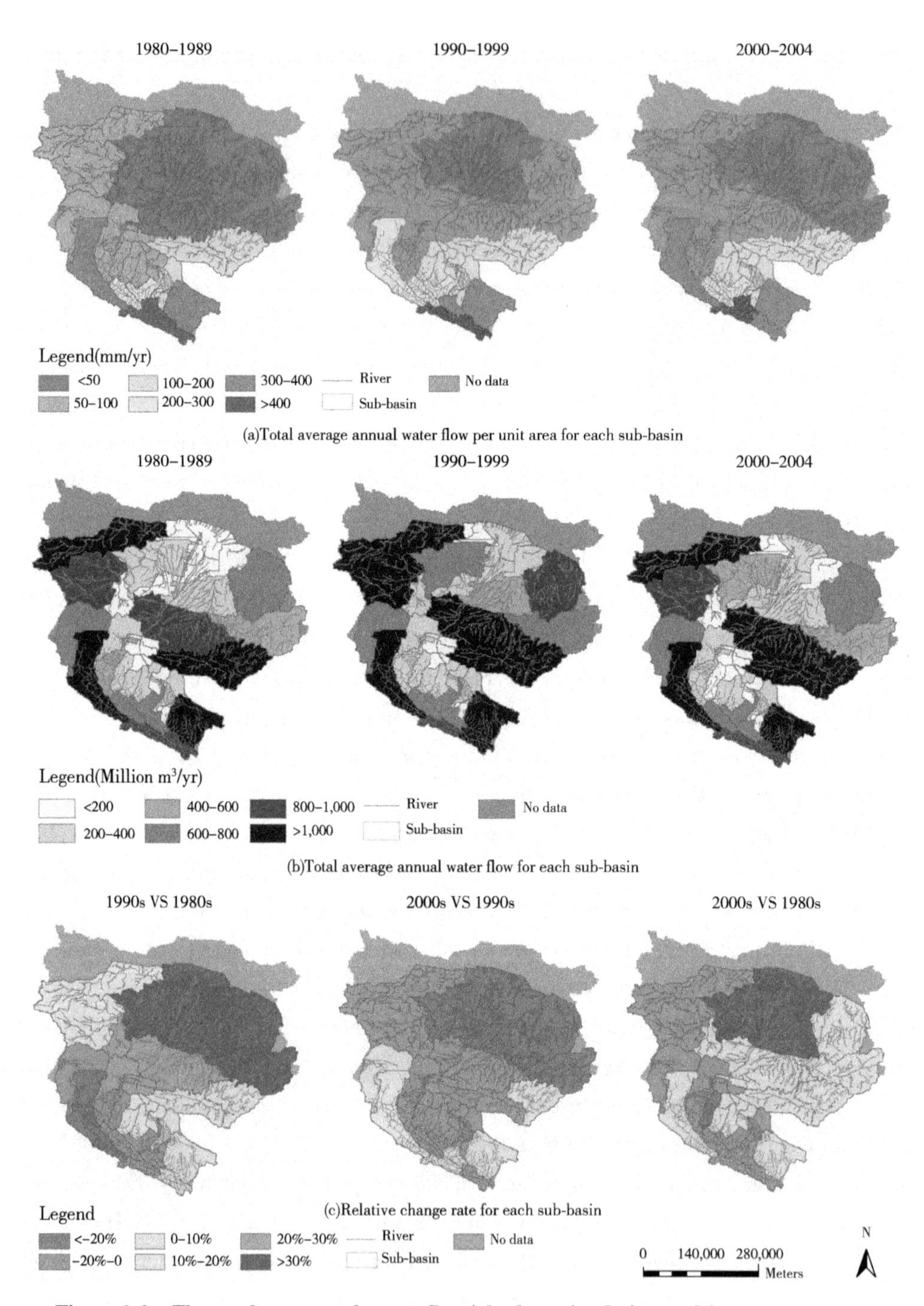

Figure 3-3 The total amount of water flow (the best simulation and long term average annual values) and its relative change rate in the Heihe River Basin

per unit area are mainly influenced by the spatial patterns of precipitation, which generally decreases from upstream to downstream. Land cover also plays a role here. Sub-basins with snow and melting water often have higher blue water flows per unit area.

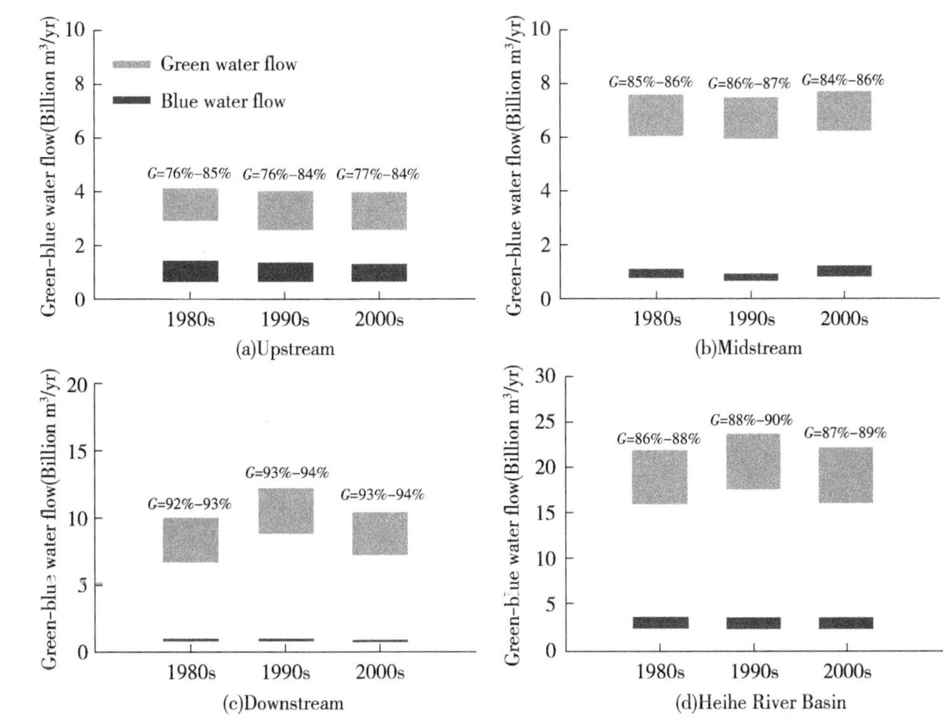

Figure 3-4 The total water flow and green water coefficients from the 1980s to the 2000s in the Heihe River Basin(*G* is green water coefficient)

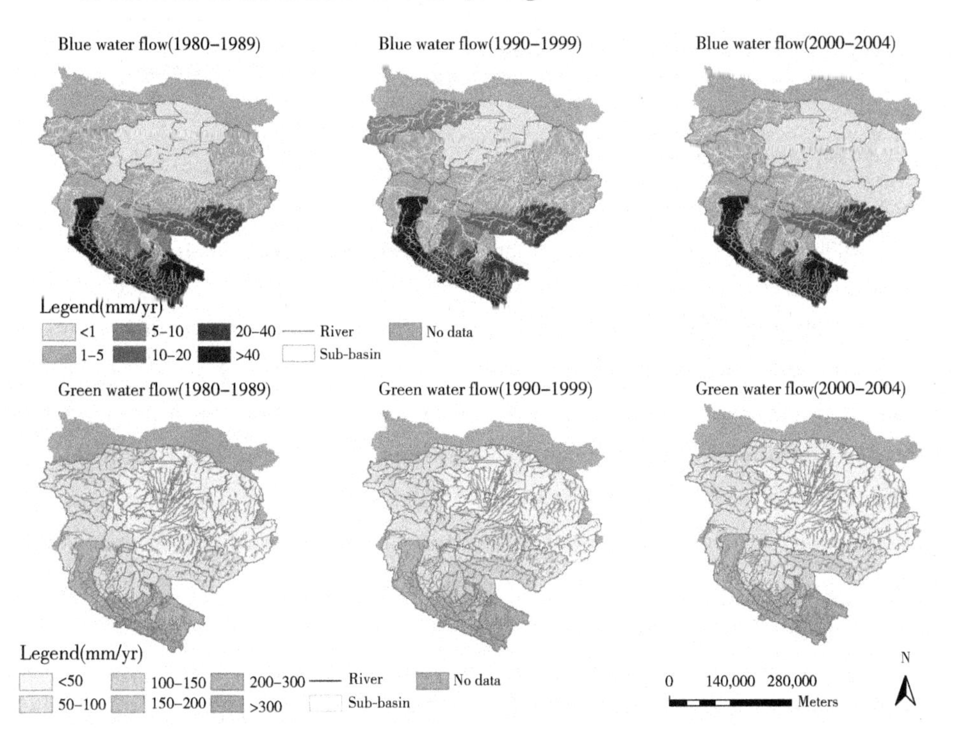

Figure 3-5 The green-blue water flows(the best simulation and long term average annual values) per unit area(mm/yr) from the 1980s to the 2000s in the Heihe River Basin

The blue water flows in the Heihe River Basin were generally high in upstream sub-basins and low in downstream sub-basins (see Figure 3-6). Two factors contribute to this spatial pattern: precipitation and land cover type. In upstream sub-basins, precipitation is generally high where snow and melting water often exist. Both the conditions result in a relatively large amount of runoff and blue water flows. In downstream sub-basins, precipitation is very low while desert is the dominant land cover. Runoff is small and hence blue water flows are low.

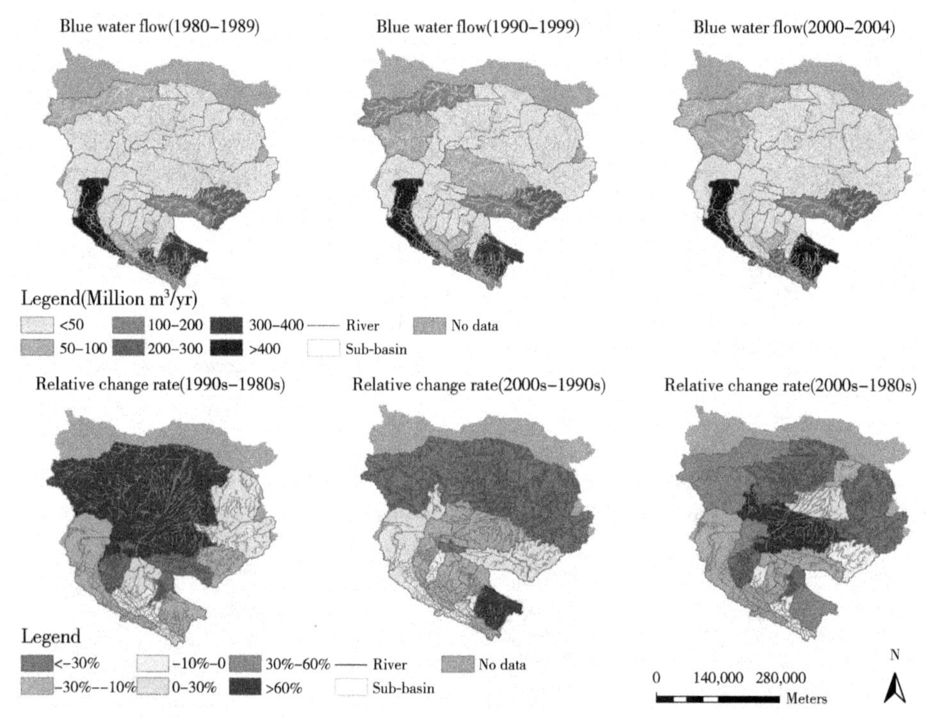

Figure 3-6 The blue water flows (the best simulation and long term average annual values) (million m³/yr) from the 1980s to the 2000s in the Heihe River Basin

It seems that, from 1980s to 1990s, blue water flows relative change rate decreased upstream and midstream and increased downstream (see Figure 3-6). However, from 1990s to 2000s, relative change rate has different trends occur with blue water flows increasing upstream but decreasing downstream. When comparing blue water flows in the 1980s with those in the 2000s, there are no clear trends of changes among regions. We can not identify a clear trend related to climate change. Climate variations in the Heihe River Basin influences precipitation and temperature, which caused the variation in blue water flows.

Green water flows are distributed more homogeneously than blue water flows among regions. In upstream sub-basins, precipitation is high, but due to the low temperature, evapotranspiration may be relatively small. In downstream sub-basins, precipitation is low, but in the desert areas, there is little runoff, or in the other words, precipitation is almost directly evaporated into the atmosphere. Besides the climatic factors and land cover, the area of sub-basins is often larger downstream than upstream. This also contributes to the more even distribution of green water flows. Furthermore, the green water flow quantity is similar to previous research. Jin and Liang (2009) studied the actual evapotranspiration of Zhangye in the Heihe River Basin, which is located at midstream and close to Zhengyi canyon. They showed that annual evapotranspiration ranged from 238 million m^3 in the1980s to 355 million m^3 in the 2000s. Our results show annual evapotranspiration of about 200 to 400 million m^3 in the above two periods (see Figure 3-7). Similar results were also estimated by Cheng et al. (2007).

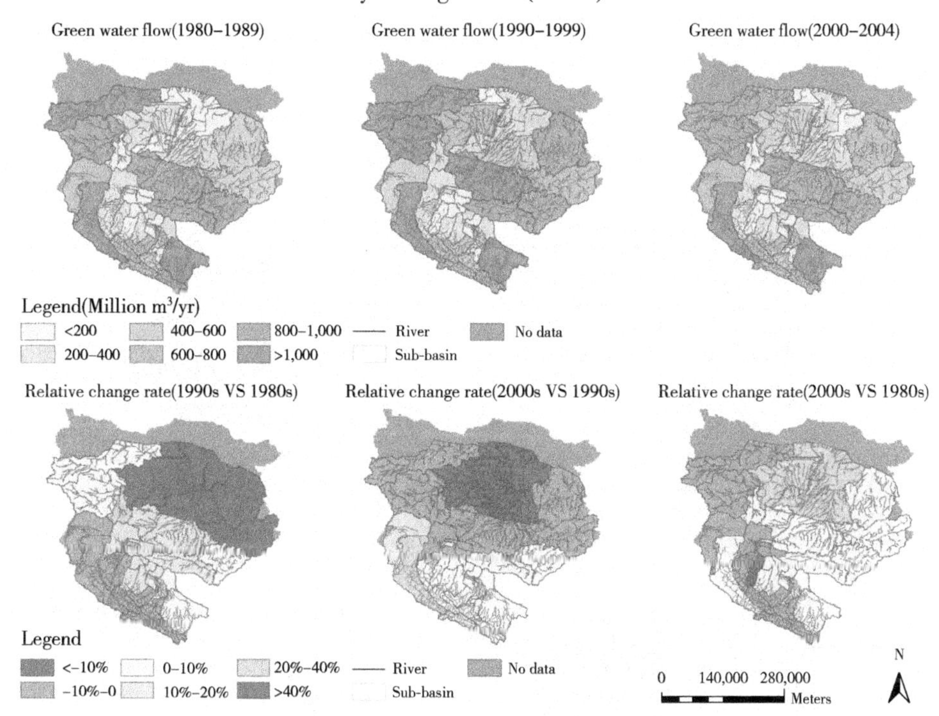

Figure 3-7 The green water flows (the best simulation and long term average annual values) (million m^3/yr) from 1980s to 2000s in the Heihe River Basin

There is no clear evidence that shows significant impacts of climate change on green water flows. In many middle and downstream sub-basins, green water flow relative change rate increased from the 1980s to the 1990s but decreased since the 1990s, while in several upstream sub-basins, green water flows decreased from the 1980s to the 1990s but increased since the 1990s (see Figure 3-7). There are no clear signals of increased or decreased green water flows with time.

Within the Heihe River Basin, the green water coefficient is relatively lower upstream and higher downstream. The green water coefficient is generally 80%-90% in upstream sub-basins, while it is generally above 90% in downstream sub-basins (Figure 3-8). The spatial distribution of green water coefficient is closely linked to land cover and geographical patterns. In upstream regions, precipitation is high at high altitude with low temperatures and evapotranspiration rates;

consequently discharge is high(Wang and Zhou,2010;Guo et al. ,2011). In particular,there is much snow and melting water upstream,which generate a large amount of runoff through melting. This is particularly obvious for one sub-basin(in dark blue) in the 1980s where the green water coefficient is even lower than 65%. This sub-basin links upstream and mid-stream as most of the upstream discharge flows through Yingluo Canyon in this sub-basin onwards to midstream and downstream(Li,2009). As a flow accumulation region,this sub-basin has the lowest green water coefficient among all sub-basins. Downstream,precipitation is low and desert is the dominant land cover. Runoff seldom occurs as precipitation mostly evaporates. Hence,the green water coefficient is extremely high. From 1980s to 2000s,the green water coefficient does not change much for most of the sub-basins(see Figure 3-8).

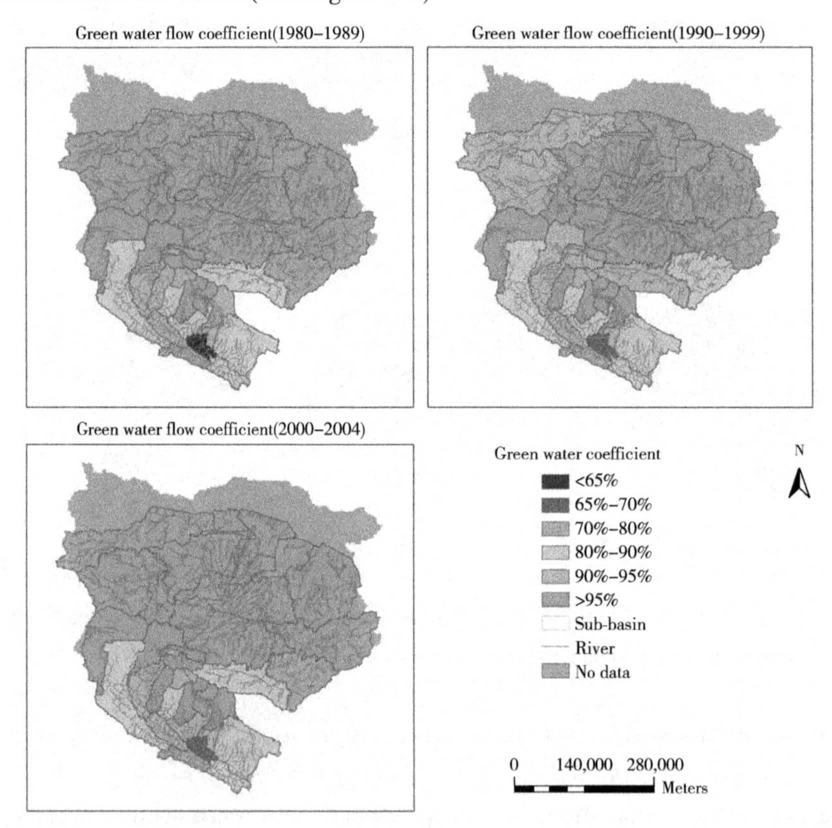

Figure 3-8 The green water coefficient (the best simulation and long term average annual values) from 1980s to 2000s in the Heihe River Basin

For the entire basin, the green water coefficient remained relatively stable over the whole time period (87%-89% in 1980s, 88%-89% in 1990s and 88%-89% in 2000s; see Figure 3-4). The green water coefficient is very high compared to previous studies on other locations, e. g. 58% in the Congo River Basin and 61% in the west of Iran(Schuol et al. ,2008; Faramarzi et al. ,2009). The high green water coefficient in the Heihe River Basin is mainly a result of the arid-and semi-arid climate conditions, which leads to low runoff and groundwater discharge but high evapotranspiration. We do not find a significant trend of change in the green water coefficient. The fluctuation of the green water coefficient also occurs upstream and midstream (see Figure 3-4) . Downstream, the green water coefficient increased from 1980s to 2000s.

3.4 Summary and conclusion

In this study, the semi-distributed SWAT model was successfully applied to quantify the green water and blue water flows for the entire Heihe River Basin. Calibration and validation at two upstream hydrological stations indicated good performance of the SWAT model in modelling hydrological processes without human intervention. The spatial and temporal distributions of blue water and green water flows were presented for the entire river basin.

Generally, green water and blue water flows per unit of area decrease from upstream to downstream. The total water flow in the Heihe River Basin has changed little during 1980-2004. Since we do not consider human intervention in the simulation, the changes are completely related to climatic factors, i. e. precipitation and temperature. Our results show variation without any clear temporal trend on total water flow in the Heihe River Basin. Instead, natural climate variability is likely the main reason for the temporal changes of water flows.

The present research on green water and blue water flows considers only natural conditions without human intervention, e. g. land use change. However, the water resources distribution in time and space has been altered by human activities midstream. This has led to significant deviations of the green water and blue water flows and transformations from what would be expected under natural conditions. Therefore, including human activities for the simulation of green-blue water flows in the Heihe River Basin is necessary, and will be the next step of our research.

This study is limited by several shortcomings. Firstly, the limited number and uneven distribution of weather and hydrological stations (see Figure 3-1) influences the accuracy of results. Only 19 weather stations and two hydrological stations were used in this study and shortage of data will influence simulation accuracy. Secondly, for now we neglect the effects of irrigation water use, land use change and operation of reservoirs. Clearly, human activities, especially the expansion of irrigated area, have already influenced the water cycling significantly in the Heihe River Basin. However, the extent of the hydrological responses to human intervention has never been assessed quantitatively. The current study provides a first step for such an assessment by quantifying the green water and blue water flows under natural conditions. Thirdly, although the two upstream discharge stations represent 85% of the annual discharge flows, they represent a small proportion of the entire region. Using only these two stations may work satisfactorily for simulating green water and blue water flows under natural situations, but more hydrological stations are still needed in the next step to study the effects of human activities on hydrological flows. Last but not least, the lack of soil moisture and actual evapotranspiration data hampers the validation of the green water flow simulations.

This study provided insights into green water and blue water flows for the entire Heihe River Basin at sub-basin level. This information is very useful for developing an overview of the actual water resources status and will provide a theoretical reference for the water resources management of the inland river basins of China.

Chapter 4 The Spatial and Temporal Dynamic Distribution Pattern of Blue-Green Water in The Heihe River Basin under the Influence of Human Activities

4.1 Research background

The impact of climate change on water availability has caused sustainability concerns around the world (Piao et al., 2010; Vörösmarty et al., 2000). With global warming, extreme weather events are now occurring more frequently, resulting in serious challenges to water supply (Vörösmarty et al., 2000, 2010; Oki and Kanae, 2006; Alley et al., 2003). Global climate change has already decreased water resources in many regions (Kundzewicz et al., 2008). In particular in arid regions, water use competition is intense among human, agriculture and ecosystems (Cheng et al., 2003; Vörösmarty et al., 2000). This influences socioeconomic sustainability and ecosystem health, and may lead to ecosystem degradation (Piao et al., 2010; Falkenmark et al., 2003; Cheng et al., 2006). Therefore, a comprehensive study on the available water resources in the context of global change is critical for an in-depth understanding of the variability of water resources for better water resources management.

In recent years, scholars have paid much attention to blue water resources assessment and management, but they have not paid sufficient attention to green water that is important for both human and ecosystems (Falkenmark, 1995; Cheng et al., 2006). Blue water is the water that is stored in rivers, lakes, aquifers and wetlands, while green water refers to the soil water from precipitation that is used for plant transpiration and soil evaporation (Falkenmark, 1995). Green water plays a critical role in producing food and maintaining natural ecosystems (Falkenmark et al., 1995, 2006). Rost et al. (2008) and Liu et al. (2009a, b) estimated green water accounted for more than 80% of consumptive water use of global crop production. Furthermore, water use in other ecosystems (e. g. grassland and forest ecosystems) is dominated by green water (Rockström, 1999; Rost et al., 2008).

Water resources in a catchment is influenced not only by climate change but also by activities of humans (Wang et al., 2005). Actually, there is interaction among climate factors, human activities and water resources. Water vapour cycling is a critical part of the climate system; and climate change will also influence the spatial and temporal variability of water resources in a watershed (Chen et al., 2007; Ren, 2011). On the one hand, human activities (e. g. land use change and irrigation) will change the discharge generation mechanism; on the other hand, human activities (e. g. greenhouse gas emissions) will trigger global or regional warming, consequently influencing water resources distribution (Siriwarden et al., 2006; Thamapark et al., 2006). The impacts of human activities and climate factors may cause a water shortage or crisis, especially in arid and semi-arid regions (Xiao et al., 2004). Therefore, assessments of the joint impacts of human activities and climate variability and trends on water resources are needed to understand green and blue water variability and sustainability of freshwater use in arid regions.

Recently, several studies have investigated the impacts on water resources of human activities, including land use change (Postel et al., 1996; Gerten et al., 2008; Jewitt et al., 2008; Liu et al., 2009), water conservancy projects (Hayashi et al., 2008; Liu et al., 2013), and irrigation (Allen et al., 1998; Wang et al., 2003). Li et al. (1998) found that human activities mainly in terms of irrigation decreased runoff in downstream locations of the Tarim River Basin. Fang et al. (2001) found that the groundwater table declined and land subsidence had happened in several cities in the northern part of China due to groundwater over-exploitation. Williams (2009) estimated that water conservancy projects caused runoff to decrease by more than 55% in the Nile River Basin.

And Wang et al. (2003) found that cropland irrigation decreased runoff and changed overall water flows in the Hexi corridor (Heihe River Basin, China). However, how human activities and climate factors synchronously impact on both green and blue water flow and their spatial distribution remains a rarely studied issue. In addition, many studies emphasize global change impacts on water resources in the future by establishing future scenarios (Xu et al., 2003; Wei et al., 2009), but few focus on the past.

In this study, we investigated the impacts of three main processes, i. e. land use change, irrigation expansion and climate variability, on the flows of green and blue water in the Heihe River Basin over the period 1986 to 2005. Based on these results, we discuss implications for future research and water management. The Heihe River Basin is the second-largest inland river basin in China. It is located in the northwest of China, and it has suffered from a serious water crisis in recent years (Cheng et al., 2003, 2006). Following the socio-economic development, water use in the midstream has increased sharply (Ma et al., 2011), and recently, human activities have changed the distribution of lakes and watershed and the inter-annual allocation of water resources in the Heihe River Basin (Xiao et al., 2004). So far, no efforts have been paid to the assessment of green water and blue water in the context of global change. We here investigate the green water and blue water dynamics and their natural and anthropogenic causes.

4.2 Methodology

The Heihe River originates in the Qilian Mountains, and discharges into the Juyanhai Lake. The area of the basin is 0.24 million km^2, with the majority located in China and a minor part in Mongolia (see Figure 4-1). The basin has an average altitude of over 1,200 m with a length of 821 km for the major channel. Three sections are often distinguished: upstream from the Qilian Mountain to the Yingluo Canyon, midstream running from the Yingluo to Zhengyi Canyon, and downstream terminating in the Juyanhai Lake (see Figure 4-1). The average annual air temperature is 2-3 ℃ in the upstream, 6-8 ℃ in the midstream, and 8-10 ℃ in the downstream area. The average annual precipitation is 200-500 mm in upstream, 120-200 mm in midstream, and < 50 mm in most downstream regions (Cheng et al., 2003). The potential evaporation ranges from 1,000 mm/yr in the upstream to 4,000 mm/yr in the downstream area (Li, 2009). Precipitation occurs mainly in summer and autumn (>70% of annual precipitation between May and August; Ma et al., 2011), whereas spring is dry with snow and ice melting (4% of total discharge; Li, 2009); there is much snow in winter (Cheng et al., 2003). The river discharge provides about 65% of the blue irrigation water in midstream regions, while groundwater provides over 90% in the downstream regions (Xiao et al., 2011). The main land cover types of the basin include desert (prevailing downstream), mountains (upstream) and oases (midstream); these three land covers together account for 98.6% of the total basin area (Cheng, 2003).

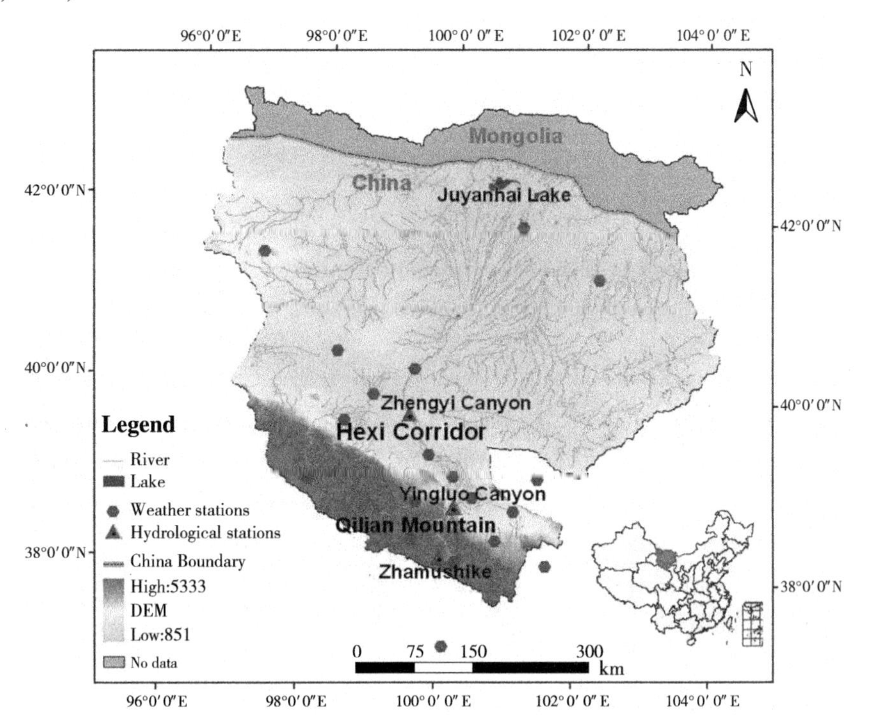

Figure 4-1 The Heihe River Basin with rivers, lakes, hydrological and weather stations indicated (The location of the Heihe River Basin in China is shown in the inset)

To study climatic and anthropogenic impacts on green water and blue water resources in the Heihe River Basin, we set up four simulation experiments as follows: scenario A fixes land use and climate conditions around 1986(land use in 1986 and climate for 1984-1986);scenario B uses land use in 1986 and climate conditions for 2004-2006;scenario C uses land use in 2005 and climate conditions for 2004-2006;scenario D assumes all crops are irrigated in addition to scenario C(see Figure 4-2). Based on these scenarios, we analyse the impacts on green water and blue water flows of climate variability (difference between B and A), land use change (difference between C and B), irrigation expansion(difference between D and C) and all factors (difference between D and A), respectively.

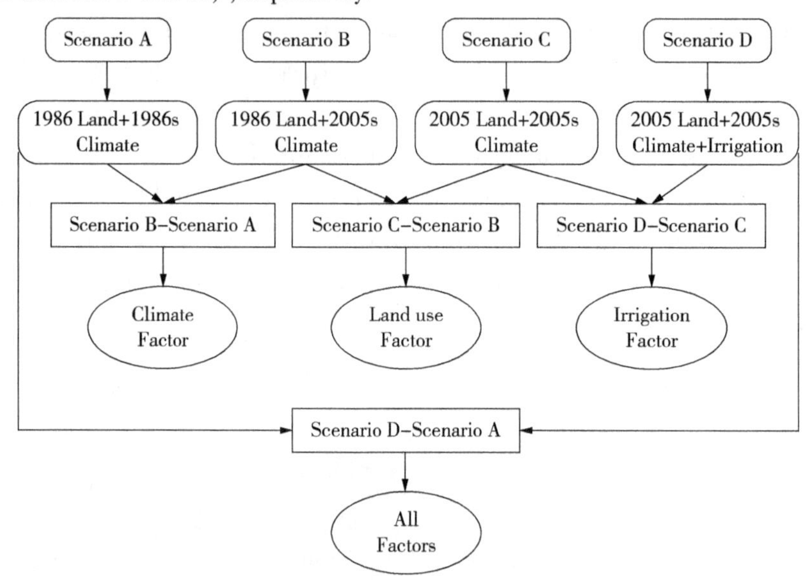

Figure 4-2 **The scenarios setup and research framework**

(1986s **is the average of** 1984 **to** 1986;2005s **is the average of** 2004 **to** 2006)

In this study, we used the Soil and Water Assessment Tool(SWAT) to simulate green water and blue water flow for the Heihe River Basin. The SWAT model is a semi-distributed water assessment model (Neitsch et al., 2004), which has been applied widely in different regions across the world(e. g. Schuol et al., 2008; Faramazi et al., 2009). We select the SWAT model for this study mainly due to two reasons. Firstly, it has been successfully used for assessments of water cycling processes under different environmental conditions(Faramazi et al., 2009; Schuol et al., 2008; Gassman et al., 2007); and secondly, it has been successfully tested to simulate hydrological processes in the Heihe River Basin (Huang et al., 2004; Li et al., 2009), including the green water and blue water flows(Zang et al., 2012).

We use the version of SWAT 2005 that works in Arcview 3. 3. The study area was separated into 309 hydrological response units(HRUs) and 32 sub-basins with information on topography, land use type, soil attributes, and management. Evapotranspiration was estimated by the Hargreaves method(Hargreaves et al., 1985), surface runoff was calculated with an SCS Curve Number(CN) method (Neitsch et al., 2004), while snowmelt was computed by an energy balance approach(Neitsch et al., 2004).

The actual evapotranspiration includes plant transpirationand soil evaporation. The SWAT model first calculates rainfall interception by plant canopy, then the maximum plant transpiration and soil evaporation using an approach similar to Ritchie(1972). The actual plant transpiration and soil evaporation are then calculated based on the soil moisture balance following Neitsch et al. (2004). The land use types influence surface runoff generation rate and evapotranspiration. In SWAT, different land cover types (urban, crop and forest) correspond to different parameters (e. g. for curve number) (Neitsch et al. , 2004). Irrigation influences the discharge in water channels and transpiration of irrigated crop. We only worked on the part of the river basin located within China and did not include the Mongolia part due to the lack of data. We use the SUFI-2 approach from the SWAT-CUP interface (Abbaspour et al. , 2007) to optimize parameters. The Nash-Sutcliffe coefficient (E_{ns}) (Nash et al. , 1970) and the coefficient of determination (R^2) were used to evaluate the goodness of the calibration and validation process.

This study is an expansion of our previous research. In Zang et al. (2012), we assessed the spatial and temporal distribution of the water flows of green and blue only under natural conditions, i. e. without considering human activities. The SWAT model was calibrated to simulate the green water and blue water flows at the whole-basin level by using climate data from 1980 to 2004 and land use data for 2000 (Zang et al. , 2012), we used the calibrated parameters derived from the previous study, and further investigated the impacts of human activities (land use change and irrigation expansion) and climate variability on the green water and blue water flows. The impacts are assessed by comparing results from different scenarios with those under natural conditions. We expand the study period to cover 1978-2005, and land use maps for 1986 and 2005 are used in addition to that for 2000 in previous research.

The actual evapotranspiration is green water flow, whereas the sum of surface runoff, lateral flows, and groundwater recharge is treated as blue water flow (Schuol et al. , 2008). To account for the relative importance of the two flows, we defined the green water coefficient (GWC) as the ratio of green water flow to the total flow (green water and blue water flows) (Liu et al. , 2009a). The relative change rate ($RCR = [(v_i - v_0)/v_0] \times 100\%$) was applied to indicate the relative change of a variable. v refers to a variable (e. g. green water flow), 0 indicates the initial time period, and i refers to the ending time period.

The data on daily climate, DEM and land use in 1986 were obtained from the Heihe Data Research Group (HDRG) (http://westdc. geodata. cn). We used climate data between 1980 and 2005 from 19 weather stations for our simulation: 7 upstream stations, 7 midstream stations, and 5 downstream stations. The irrigation area, irrigation depth, and irrigation parameters that need input the SWAT model were obtained from published literature (Ge et al. , 2011; Wang et al. , 2012) and the Ministry of Water Resources irrigation test web site. The irrigation districts data were obtained from the HDRG (http://westdc. geodata. cn), and the total irrigation area in the river basin is 1. 88 million ha. In scenario D, we assume that all cropland within the irrigation districts is irrigated. The 1 km land use data for 1985 and 2005 were obtained from the institute of Geographic Science and Nature Resources Research, Chinese Academy of Sciences (CAS). The soil data were obtained from the Harmonized World Soil Database (HWSD) (http://www.iiasa.ac.at) with a spatial resolution of about 1 km. This dataset includes 63 soil types for the Heihe River Basin, and for each soil type soil parameters for two soil layers are available, i.e. 0-30 cm and 30-100 cm.

4.3 The influence of human activities on spatial and temporal distribution of blue-green water

4.3.1 The overall change of blue-green water flow in the Heihe River Basin under the influence of human activities

Impacts of land use change are assessed by comparing difference of results from land use in 1986 and 2005 (holding all other factors unchanged). According to our model simulation results (Table 4-1), at the river basin level, land use change has resulted in an increase of blue water flow by 206 million m^3 and a concurrent increase in green water flow by the same amount. These changes are simulated in particular for the midstream and a part of the downstream region (see Figure 4-4 and Figure 4-5). The relative change rate in the sub-basins there was more than 50% (Figure 4-3). Irrigation water use and urban land use are the two main reasons that caused the hydrological variability in the SWAT model (Neitsch et al., 2004). In this section, the main reason for these changes was, that urbanization expanded fast in these sub-basins (Figure 4-6), accelerating surface runoff production. In our simulations, the total green water and blue water flows did not change in response to the land use changes, but the green water coefficient is found to have decreased from 81%-90% to 71%-75%, in particular in the middle part of midstream (see Figure 4-7).

Table 4-1 The variability of blue water flow, green water flow and total green/blue water flows by different factors influence of Heihe River Basin

Project	Climate change	Land use	Irrigation	Comprehensive change
Blue water flow	1.46	2.06	-0.66	2.86
Green water flow	4.69	-2.06	0.66	3.29
Blue and green water total	6.15	0	0	6.15

4.3.2 The change of blue water flow in the Heihe River Basin under different situations

At the river basin level, irrigation expansion resulted in a decrease of blue water flow by 66 million m^3, according to our model (see Table 4-1). Blue water flow decreased in particular in midstream regions (see Figure 4-3), where a large area of agriculture with many irrigation farmlands exists (see Figure 4-4). In an earlier study, irrigation expansion was shown to require a large amount of water from rivers and groundwater (Wang et al., 2003). Green water flow has

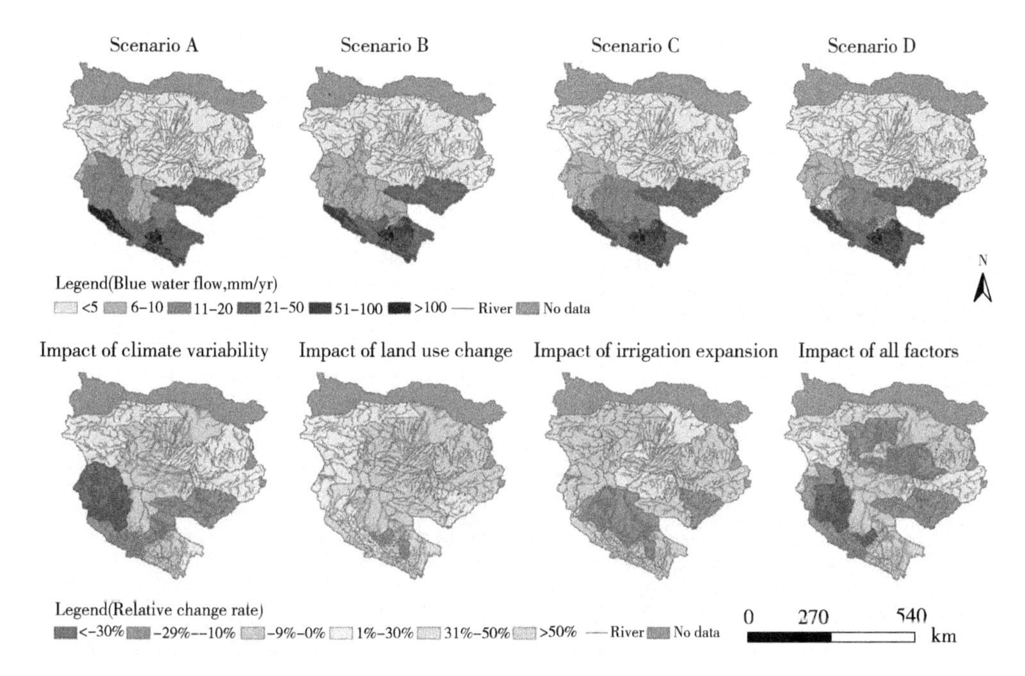

Figure 4-3 The blue water flow in the Heihe River Basin in different scenarios. The impacts are assessed with relative change rate. The impacts of climate variability, land use change and irrigation expansion are assessed by comparing results between Scenario B and Scenario A, Scenario C and Scenario B, and Scenario D and Scenario C, respectively. The impacts of all factors together are assessed by comparing results between Scenario D and Scenario A

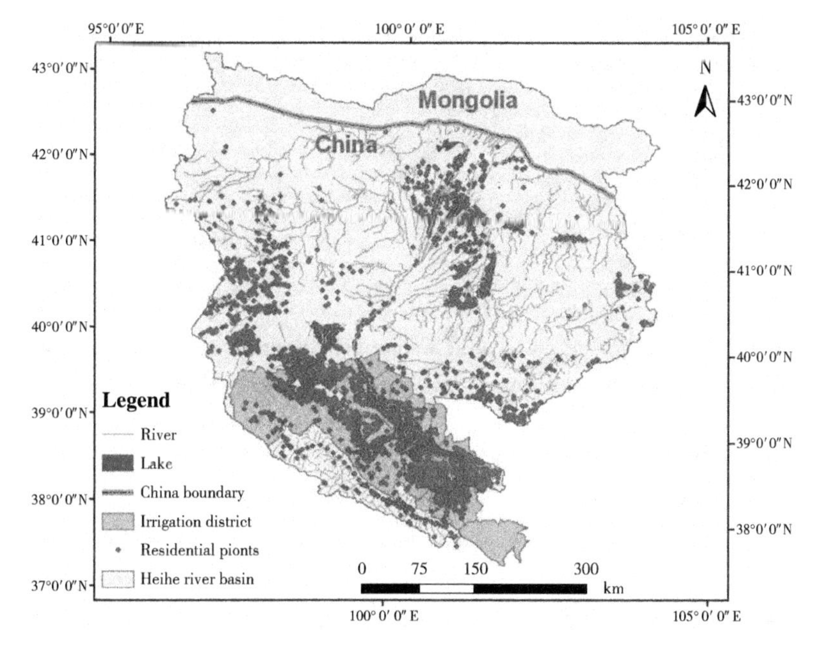

Figure 4-4 The distribution of irrigation district and residential points in the Heihe River Basin.

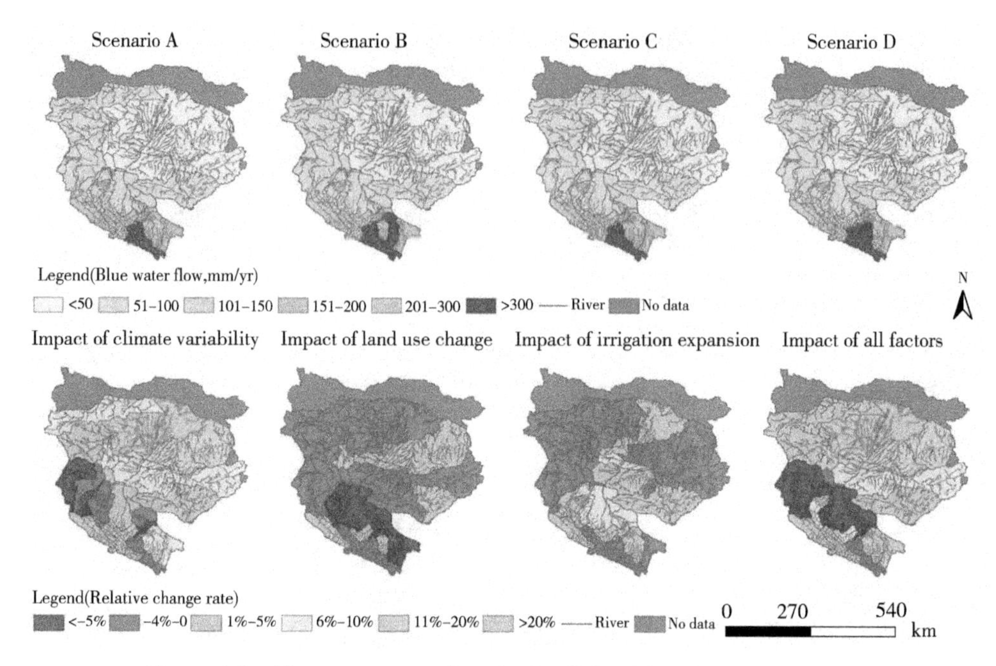

Figure 4-5 The green water flow in the Heihe River Basin in different scenarios (Details as in Figure 4-3)

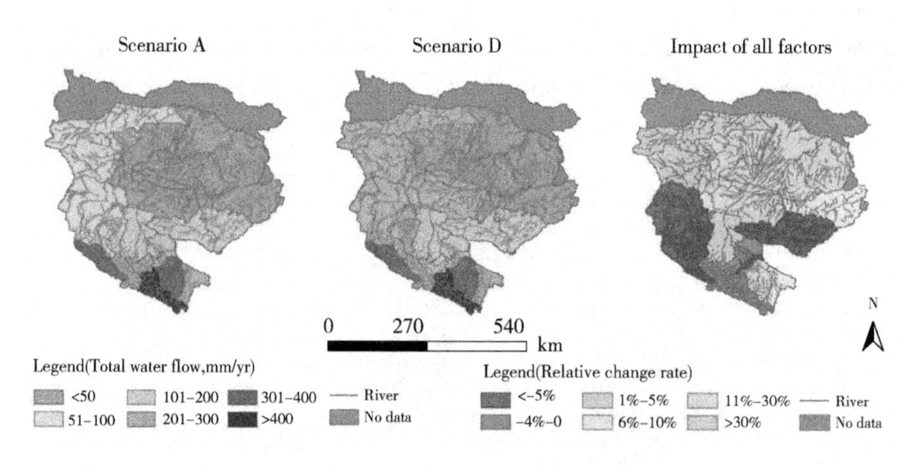

Figure 4-6 Impacts of all factors (human activities and climate variability) on the total water flows in the Heihe River Basin

increased by the same amount due to farmland irrigation (see Figure 4-5), because, compared to rainfed agriculture, irrigated agriculture consumed more water and thus increased actual evapotranspiration. As in the case of the isolated land use change effect, the total green water and blue water flows did not change at the river basin level due to irrigation expansion, and the green water coefficient has increased from 71%-75% to 81%-90% in particular in the eastern part of the midstream area (see Figure 4-7).

4.3.3 Changes of green water flow and green water coefficient in the Heihe River Ba-

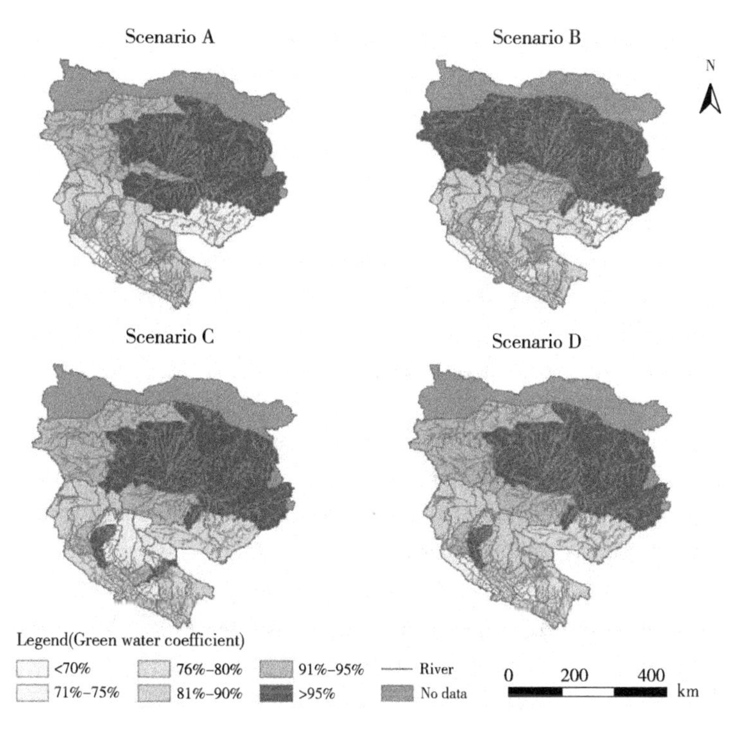

Figure 4-7　The green water coefficient in the Heihe River Basin in different scenarios

sin under different situations

Climate variability, assessed as the difference between the mid-1980s and the mid-2000s, has increased both blue water flow and green water flow by 146 million m^3 and 469 million m^3, respectively (see Table 4-1), with little change in the green water coefficient (87%-88%). Spatially, although climate variability has led to an increase in blue water and green water flows in most sub-basins, we can also find a clear decreasing trend of both flows in the western part of the midstream. The decrease was a result of a lower precipitation in the western midstream area, where precipitation has decreased significantly at $p < 0.10$ level from 1980 to 2005 (see Figure 4-8). Precipitation had increased in downstream areas; hence, blue water and green water flows have increased there (see Figure 4-8).

In response to climate, land use and irrigation change, blue water flow has increased by 286 million m^3 in the entire river basin (see Table 4-1). The spatial distribution of the changes in blue water flow varies largely, with decreases in western sub-basins, but increases in eastern sub-basins in midstream areas. The relative change rate of several midstream sub-basins exceeded 30% (see Figure 4-3). The change patterns were found to be largely influenced by climate variability; for example, in the western part of midstream areas, precipitation showed a decreasing trend (see Figure 4-8); hence, both blue water and green water flows decreased, though also influenced by other factors. As shown above, the accelerated urbanization and farmland irriga-

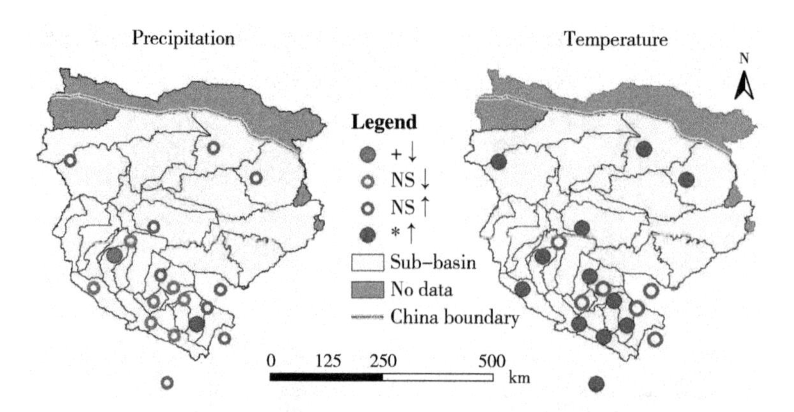

Figure 4-8 The variability of precipitation and temperature in the Heihe River Basin from 1980 to 2005(↑ indicates increasing trend; ↓ decreasing trend, * significant at $p < 0.05$; + significant at $p < 0.10$; NS, not significant)

tion are the other main reason that caused blue water flow variability, the increase of blue water flow probably influenced by urbanization development, and the decrease by irrigation expansion (Wang et al. ,2003; Maet al. ,2008). Therefore, land use change contributed most to this increase, followed by the contribution from climate variability(see Table 4-1).

Green water flow showed an increasing trend with a 329 million m^3 higher flow over the entire river basin, mainly due to climate variability but also due to irrigation expansion (see Table 4-1). Decreases(>30%) are found to prevail in most midstream sub-basins, while increases were simulated for most of sub-basins downstream (see Figure 4-5). The total water flows increased by 615 million m^3, caused almost exclusively by climate variability. Both land use change and irrigation expansion did not alter the amount of total water flows; instead, they influenced the allocation of water into green water or blue water flows.

4.4 Discussion and summary

In this study, we applied the SWAT model to analyze the impacts of human activities and climate variability on green water and blue water flows for an arid river basin in a spatially explicit way. We choose two time periods(around 1986 and around 2005) to analyze the green water and blue water flow variability because these two periods reflect the sharp socio-economic changes in the past decades. Land use change and irrigation were used as indicators for human activities. The land use change was found to be a main factor that influences water resources variability. Between 1986 and 2005, the urban area has increased by 47%, and the irrigated land had increased by 27%.

From the different scenarios of blue water and green water changes can be seen in the human activities under the influence of blue water and green water there is a certain relationship between the transformation. Under fixed climatic conditions, the change in land use increased the flow of blue water by 206 million m^3 and the flow of green water by 206 million m^3. Irrigation reduced the flow of blue water by 0. 66 million m^3, and the increase in green water was 0. 66 million m^3. Through this result can be seen in the case of constant water resources, blue water and green water in the human activities under the influence of a certain conversion. In the blue water to green water conversion process as follows: precipitation→blue water→water and plant transpiration to form green water flow; and green water flow to the blue water flow conversion process as follows: precipitation→green water→atmospheric condensation again to form precipitation→blue water. From these two processes, it can be seen that the conversion process of blue water flow to green water flow is more direct, and the conversion of green water flow to blue water flow needs to be completed by secondary precipitation. This paper simply lists the blue-green water conversion process, but the specific process of blue-green water conversion process and the mechanism is not clear, so to strengthen the conversion of blue water and green water research, can effectively improve the water resources usage efficiency of the arid and semi-arid areas.

Through the analysis of the spatial and temporal distribution pattern of blue-green water in the Heihe River Basin under the influence of human activities and the analysis of the main reasons, we draw the following conclusions:

(1) The change of green water flow in the Heihe River Basin under the comprehensive scenario is mainly manifested in the decrease of the sub-basin. From the comprehensive situation, the Heihe River Basin blue water flow and no obvious changes in the law. However, under the dual effects of climate fluctuation and human activities, the total amount of blue-green water in the Heihe River Basin increased obviously in the sub-basin and the southeastern sub-basin in the middle reaches of the lower reaches. In the sub-basin and the southeast watersheds are reduced.

(2) Under the climate change scenario, the blue water flow in the Heihe River Basin has been significantly reduced in the middle reaches of the basin, and the fluctuation of the climate has increased the blue water flow in the Heihe River Basin by 286 million m^3 over the past 20 years (see Table 4-1). In the case of land use change, the blue water flow in the Heihe River Basin increased significantly in the middle reaches of the sub-basin, mainly due to the accelerated process of urbanization in the region. In irrigation scenarios, irrigation in irrigation areas requires a large amount of water from the river or shallow groundwater, which can cause blue water to decrease in the middle reaches of the sub-basin.

(3) Under the climate change scenario, the change of precipitation and temperature from 1980 to 2005 resulted in a significant decrease in the sub-watershed in the western part of the middle reaches of the Heihe River Basin, and a significant increase in the sub-watershed in the north of the lower reaches. In the land use change scenario, the Heihe River Basin in the middle reaches of some sub-basin reduction is more obvious. This is mainly due to the region in the past 20 years to accelerate the process of urbanization. In the irrigation scenario, the irrigation of

farmland increased the source of evapotranspiration, resulting in a significant increase in the green water flow in the middle and eastern irrigation areas of the middle reaches of the Heihe River Basin.

(4) Climate change scenarios, the Heihe River Basin green water in the lower reaches of the northwest region increased. In the case of land use change, the green water coefficient is decreasing in the sub-basin of the middle and eastern region in the middle reaches. In the irrigation scenario, the green water coefficient in the middle reaches of the Heihe River irrigation area showed a significant increase.

In this paper, the changes of blue-green water in the Heihe River Basin under the conditions of climate change scenarios, land use scenarios and irrigation scenarios are used to interpret the changing laws of blue-green water in the Heihe River Basin under the influence of climate change and human activities. This study provides a basic reference for comprehensive understanding of the distribution and change of water resources in the Heihe River Basin, and can provide the necessary reference and theoretical guidance for further research and scientific management of the watershed.

Chapter 5 Study on Spatial and Temporal Differences of Blue-Green Water in The Heihe River Basin in Typical Years

5.1 Determination of typical years

This study aims to provide a theoretical study on the spatial and temporal evolution mechanism of blue-green water under the influence of climate change by analyzing the temporal and spatial characteristics of blue-green water in the Heihe River Basin in different typical years (drought years, wet years and plain water years). And to provide scientific guidance for water resources management in the inland river basins in Northwest China. In order to avoid the error of single index calculation, this study uses standard precipitation index (SPI) and precipitation anomaly index (M) to determine the typical dry, wet and flat years.

5.1.1 Standardized precipitation index (SPI)

The standardized precipitation index is one of the indices that characterize the probability of precipitation in a given period, which was proposed by McKee et al. In assessing drought conditions in Colorado, USA (Seiler et al., 2002; et al., 2012). This indicator has the characteristics of multi-time scale application, making it possible to use the same drought index to reflect the different time scales and different aspects of water resources status, and thus widely used (Yuan Wenping et al., 2004). It is difficult to compare precipitation with different precipitation in different time and time, and the precipitation distribution is a kind of skew distribution, which is not a normal distribution. Therefore, Γ distribution is used in precipitation analysis. Probability to describe the change in precipitation, and then by normalization of the SPI value obtained.

Assuming that the precipitation of a certain period is x, the probability density function of its Γ distribution is:

$$f(x) = \frac{1}{\beta^{\alpha}\Gamma(\alpha)}x^{\alpha-1}e^{-x/\beta} \quad (x > 0) \tag{5-1}$$

$$\Gamma(\alpha) = \int_0^{\infty} x^{\alpha-1}e^{-x}dx \tag{5-2}$$

Where, α is the shape parameter; β is the scale parameter; x is the precipitation, and $\Gamma(\alpha)$ is the gamma function.

The best α, β estimates can be obtained using the maximum likelihood estimation method:

$$\hat{\alpha} = \frac{1 + \sqrt{1 + 4A/3}}{4A} \tag{5-3}$$

$$\hat{\beta} = \frac{\bar{x}}{\hat{\alpha}} \tag{5-4}$$

$$A = \ln(\bar{x}) - \frac{\sum_{i=1}^{n}(x)}{n} \tag{5-5}$$

Where, n is the length of the calculated sequence.

Thus, the cumulative probability for a given time scale can be obtained as follows:

$$g(x) = \int_0^x f(x)\,dx = \frac{1}{\hat{\beta}^{\hat{a}}\Gamma(\hat{\alpha})}\int_0^x x^{\alpha-1}e^{-x/\beta}dx \tag{5-6}$$

Let $t = x/\hat{\beta}$, the above equation be solved as an incomplete gamma equation:

$$g(x) = \frac{1}{\Gamma(\hat{\alpha})}\int_0^x t^{\hat{\alpha}-1}e^{-t}dt \tag{5-7}$$

Since there is no case where x is 0 in the gamma equation, the actual precipitation will be zero. So the cumulative probability can be expressed as:

$$H(x) = q + (1 - q)g(x) \tag{5-8}$$

Where, q is the probability that the precipitation is zero.

If m represents the amount of precipitation in the precipitation time series of 0, then $q = m/n$. The cumulative probability $H(x)$ can be converted to a standard normal distribution function by:

When $0 < H(x) \leqslant 0.5$:

$$SPI = -\left(t - \frac{c_0 + c_1 t + c_2 t^2}{1 + d_1 t + d_2 t^2 + d_3 t^3}\right) \tag{5-9}$$

$$t = \sqrt{\ln\frac{1}{H^2(x)}} \tag{5-10}$$

When $0.5 < H(x) < 1$:

$$SPI = t - \frac{c_0 + c_1 t + c_2 t^2}{1 + d_1 t + d_2 t^2 + d_3 t^3} \tag{5-11}$$

$$t = \sqrt{\ln\frac{1}{[1 - H(x)]^2}} \tag{5-12}$$

Where, $c_0 = 2.515,517$; $c_1 = 0.802,853$; $c_2 = 0.010,328$; $d_1 = 1.432,788$; $d_2 = 0.189,269$; $d_3 = 0.001,308$.

According to the above formula, can be obtained SPI, the level of drought and drought in Table 5-1.

5.1.2 Percentage of precipitation anomaly (M)

The percentage of precipitation anomaly (M) reflects the degree of deviation between the precipitation and the average state of the same period. There are different mean precipitation in different periods in different regions. Therefore, it is a relative index with time and space contrast. In the daily operations of the meteorological department, the percentage of precipitation anomalies are often used as indicators of drought and flooding(Ju Xiaosheng et al. ,1997). The calculation method is:

Table 5-1 Grades of flood/drought based on the standard precipitation index and anomalous percentage of precipitation

Percentage of precipitation anomaly(*M*)	Standardized precipitation index(*SPI*)	Drought and flood rank
< -75%	< -1.96	Extreme drought
-75% ~ -50%	-1.96 ~ -1.48	Serious drought
-50% ~ -25%	-1.48 ~ -1.0	Medium drought
-25% ~25%	-1.0 ~ 1.0	Normal
25% ~50%	1.0 ~ 1.48	Medium moist
50% ~75%	1.48 ~ 1.96	Severe wetting
>75%	>1.96	Extremely wet

Note: This information is derived from the literature(Yuan Wenping,2004; Ju Xiaosheng,1997).

$$M = \frac{p - \bar{p}}{\bar{p}} \times 100\% \tag{5-13}$$

Where, p for a certain period of precipitation, in this paper, respectively, the whole basin and the upper and lower reaches of a year precipitation. In this paper, the average rainfall in the whole basin and the upper and lower reaches of 51 years are 51 years. Table 5-1 shows the level of drought and floods calculated from the percentage of precipitation anomalies.

5.2 Data sources

The data used in this study mainly include the precipitation data and the distribution of precipitation sites, the Heihe River Data Research Group from the Heihe major research project of the National Natural Science Foundation of China, and 19 sites for the precipitation data from 1960 to 2010. The Tyson polygon method is used to calculate the total precipitation in the whole basin and the middle and lower reaches of the region, and the difference fill method is used to complete the missing data(Sluiter, 2009). The surface precipitation is calculated by the point precipitation data combined with the Tyson polygon area represented by the site, and the surface precipitation is calculated for the whole basin and the middle and lower reaches. According to the standardized precipitation index formula, the *SPI* is calculated by programming the Excel.

The blue water and green water data in this study are the simulation results of the parameters of the SWAT model. The Heihe River Basin is divided into 32 sub-basins and 309 hydrological response units according to Chapter 4. Simulation time from 1958 to 2010, due to the first two years as a warm-up period, so the use of data for the study time from 1960 to 2010. The model is defined and validated using the Nash coefficient(E_{ns}) and the coefficient of determination(R^2). The rate of the model is from 1979 to 1987, and the validation period of the model is from 1990 to 2004. The model was determined and validated. The Zamushk and the Yingxiao

Gorge in the upper reaches of the Heihe River Basin were selected and the model was determined and validated using the monthly data of the two sites. These two sites control more than 85% of the Heihe River runoff, so the two sites to select the whole basin of the rate and verification. Through the analysis and verification of the model, the Nash coefficient and the coefficient of determination are 0.88 and 0.90 respectively, which indicates that the model simulation results used in this study are highly reliable. Blue green water and green water coefficient calculation see 2.2.7.

5.3 Spatial and temporal differences of blue-green water in different typical years

5.3.1 Typical wet and dry years in the Heihe River Basin

By calculating the standard precipitation index(*SPI*) and the percentage of precipitation anomalies(*M*) in the whole basin and the upper and lower reaches of the Heihe River, it is found that the typical years in the upper and lower reaches are not consistent(see Table 5-3). Such as extreme drought in the middle reaches of 1965, but in the upper reaches and downstream, in normal years; in the case of moderate wetlands in 1983, there were serious droughts in the lower reaches; similar cases were 1970 and 2009(see Table 5-3). Between 1960 and 2010, the upper reaches of the Heihe River Basin were typical of drought only in 1978, were typical of the wet only 1998. We selected two indicators in the normal year in 1984 as the level of water. At the same time, it can be seen from Table 5-2 that the precipitation in the dry years(1978) is very low, not only significantly lower than the wet years(1998), but also significantly lower than the average of the plain water(1984) and 51 years. This shows that the calculation of *SPI* and *M* values in this study is reliable. Therefore, the typical dry, wet and flat years selected in this paper are 1978, 1998 and 1984 respectively.

Table 5-2 Precipitation and temperature of Heihe River Basin and at up-mid-down streams from year 1960 to 2010

Years	Precipitation (mm, whole basin)	Precipitation (mm, upstream)	Precipitation (mm, mid-stream)	Precipitation (mm, down-stream)	Temperature (℃, whole basin)	Temperature (℃, upstream)	Temperature (℃, midstream)	Temperature (℃, downstream)
1960	161.83	326.34	163.30	37.60	5.26	2.99	6.79	7.85
1961	146.61	273.11	82.70	24.20	5.19	2.70	6.76	8.03
1962	146.44	308.72	65.70	19.00	5.05	2.64	6.65	7.89
1963	165.80	329.13	118.60	24.00	5.52	3.08	6.93	8.22
1964	180.13	351.74	108.60	74.40	5.31	3.02	6.84	7.90
1965	150.02	314.75	58.30	36.50	5.42	2.98	6.87	8.16
1966	152.40	303.47	103.80	35.40	5.42	3.07	6.86	8.02
1967	168.41	366.17	123.90	32.30	4.70	2.61	6.47	7.38
1968	150.19	343.58	133.00	29.50	4.97	2.89	6.66	7.45
1969	172.98	376.81	148.10	83.60	5.25	3.05	6.74	7.71
1970	144.15	226.23	129.40	23.80	5.01	2.73	6.62	7.73
1971	176.68	336.19	99.30	51.30	5.25	2.80	6.70	7.81

Continued to Table 5-2

Years	Precipitation (mm, whole basin)	Precipitation (mm, upstream)	Precipitation (mm, midstream)	Precipitation (mm, downstream)	Temperature (℃, whole basin)	Temperature (℃, upstream)	Temperature (℃, midstream)	Temperature (℃, downstream)
1972	160. 88	332. 87	87. 80	14. 60	5. 32	2. 98	6. 61	7. 79
1973	152. 31	318. 44	99. 90	66. 70	5. 43	3. 02	6. 87	8. 03
1974	159. 40	377. 57	124. 40	26. 80	4. 73	2. 33	6. 13	7. 20
1975	173. 40	386. 78	122. 00	43. 20	5. 08	2. 56	6. 52	7. 81
1976	156. 36	384. 91	78. 60	25. 90	4. 48	1. 98	5. 85	7. 10
1977	186. 10	336. 1	134. 20	44. 10	4. 90	2. 12	6. 36	7. 87
1978	143. 98	293. 86	76. 70	30. 30	5. 52	3. 06	6. 97	8. 17
1979	230. 29	396. 53	184. 30	65. 80	5. 28	2. 68	6. 57	8. 00
1980	170. 50	365. 35	85. 90	25. 80	5. 30	2. 73	6. 67	8. 07
1981	220. 09	452. 67	94. 20	49. 60	5. 09	2. 77	6. 37	7. 49
1982	191. 53	404. 05	139. 40	20. 00	5. 59	2. 90	6. 98	8. 54
1983	225. 12	492. 74	176. 20	10. 90	4. 98	2. 33	6. 33	7. 94
1984	179. 47	356. 12	125. 90	36. 30	4. 52	2. 11	5. 86	6. 93
1985	187. 64	372. 47	77. 70	26. 70	5. 12	2. 74	6. 48	7. 63
1986	174. 30	391. 35	77. 50	18. 60	5. 11	2. 52	6. 54	7. 86
1987	197. 38	364. 2	145. 50	15. 70	6. 05	3. 50	7. 38	8. 82
1988	227. 73	490. 71	153. 30	23. 90	5. 47	2. 99	6. 79	8. 01
1989	211. 72	485. 88	103. 70	15. 10	5. 54	2. 95	6. 89	8. 40
1990	186. 54	421. 58	126. 10	51. 30	5. 92	3. 42	7. 28	8. 75
1991	157. 34	349. 37	92. 10	52. 90	5. 77	3. 22	7. 13	8. 56
1992	192. 31	357. 67	139. 10	40. 90	5. 36	2. 95	6. 72	8. 07
1993	231. 42	438. 43	107. 40	52. 20	5. 32	3. 09	6. 55	7. 84
1994	180. 11	421. 54	121. 60	61. 30	6. 04	3. 57	7. 41	8. 74
1995	199. 04	399. 65	123. 80	89. 50	5. 35	2. 70	6. 67	8. 22
1996	202. 77	428. 49	152. 00	24. 60	5. 34	2. 77	6. 64	8. 02
1997	158. 04	368. 06	70. 50	51. 40	6. 08	3. 61	7. 47	8. 90
1998	231. 87	573. 6	154. 00	52. 30	6. 61	4. 05	8. 04	9. 28
1999	201. 59	417. 21	120. 30	58. 00	6. 58	4. 04	8. 02	9. 27
2000	193. 49	358. 32	105. 80	29. 60	5. 86	3. 40	7. 25	8. 58
2001	168. 32	350. 46	129. 70	21. 00	6. 32	3. 93	7. 59	9. 04
2002	205. 39	373. 3	131. 50	33. 10	6. 48	3. 95	7. 80	9. 35
2003	220. 04	547. 14	121. 50	37. 60	6. 02	3. 49	7. 44	8. 61
2004	186. 73	349. 65	121. 00	33. 80	6. 26	3. 73	7. 64	9. 05
2005	210. 71	426. 74	133. 70	29. 30	5. 92	3. 45	7. 10	8. 43
2006	191. 36	412. 76	91. 30	31. 20	6. 11	3. 60	7. 23	8. 51
2007	231. 81	382. 82	165. 20	30. 70	6. 08	3. 56	7. 19	8. 57
2008	215. 73	495. 28	139. 30	65. 20	5. 82	3. 26	7. 00	8. 51
2009	217. 20	451. 65	100. 00	10. 60	6. 15	3. 59	7. 29	8. 57
2010	208. 51	331. 2	124. 20	27. 50	6. 07	3. 61	7. 14	8. 49
Mean value	185. 38	382. 62	117. 49	37. 56	5. 52	3. 05	6. 90	8. 18

Table 5-3 Typical reference years of the Heihe River Basin in the entire river basin and at up-mid-down streams from 1961 to 2009

Years	The whole basin			Upstream			Midstream			Downstream		
	SPI	*M*	grade	*SPI*	*M*	grade	*SPI*	*M*	grade	*SPI*	*M*	grade
1961	-1.77	-56.15	Severe drought	-1.57	-38.30	Severe drought						
1964										1.57	60.26	Severe wetting
1965	-1.43	-52.35	Severe drought				-2.07	-76.43	Extreme drought			
1970	1.02	36.53	Medium drought				-1.87	51.34	Severe drought			
1972	-1.30	-40.07	Medium drought							-1.49	-60.54	Severe drought
1978	-1.89	-56.53	Severe drought	-1.17	-35.46	Medium drought	-1.50	-43.44	Severe drought	-1.97	-60.54	Severe drought
1979	1.20	29.57	Medium moist				1.97	58.28	Severe wetting			
1983	1.43	25.71	Medium moist							-1.77	-71.35	Severe drought
1984	-0.14	-5.62	normal	0.10	5.38	normal	0.34	8.45	normal	-0.45	-21.83	normal
1995										1.77	54.73	Severe wetting
1998	2.07	57.81	Severe wetting	2.07	58.14	Severe wetting	1.50	58.13	Severe wetting	1.04	36.53	Medium moist
2003	1.57	51.23	Severe wetting	1.77	63.58	Severe wetting						
2009										-1.97	-58.07	Severe drought

5.3.2 Temporal and spatial differences of blue water and green water in typical years

The total depth of blue-green water in the Heihe River Basin ranges from upstream to downstream, and the total depth of blue-green water (1978, 1984 and 1998) decreases from380.46 mm in the upstream to 122.25 mm in the middle reaches, and then to the downstream average of 56.32 mm(calculated from Table 5-4 to Table 5-6). The total depth of blue-green water increased significantly from dry years(1978) to wet years(see Figure 5-1(a))from dry years to wet years, and the total depth of blue-green water increased from 150.06 mm in dry years to wet years of 231.31 mm(calculated from Table 5-4, Table 5-6). At the same time, significant changes in the total depth of the year are mainly concentrated in the upper and middle reaches of the southeastern region, while the downstream changes are not obvious (see Figure 5-1(a)).

Table 5-4 Green-blue water depth of the upstream sub-basins in the Heihe River Basin in typical reference years

Sub-watershed code	1978		1984		1998	
	Blue water	Green water	Blue water	Green water	Blue water	Green water
24	28.95	158.6	75.88	196.02	139.03	256.89
23	48.10	217.67	49.26	213.47	75.23	279.77
31	118.85	225.71	100.59	322.93	209.04	383.97
32	89.16	232.53	91.85	278.33	47.56	354.52
21	114.72	229.84	95.12	327.96	205.63	387.42
22	51.26	280.69	69.84	332.3	159.03	400.51
Average value	75.17	224.17	80.42	278.50	139.25	343.85

Table 5-5 Green-blue water depth in the sub-basin of mid-stream of the Heihe River Basin in typical reference years

Sub-watershed code	1978		1984		1998	
	Blue water	Green water	Blue water	Green water	Blue water	Green water
25	4.7	43.56	6.42	58.47	4.57	53.01
13	5.2	56	6.73	58.17	8.22	78.26
11	5.13	43.21	7.04	57.85	5.02	52.49
12	3	72.04	3.89	61.01	5.04	83.45
14	6.35	54.96	7.91	56.99	9.9	76.49
19	7.31	54.03	8.91	56	11.32	74.99
28	7.41	80.7	2.12	64.51	8.77	106.01
18	2.45	21.36	3.46	57.49	13.12	101.42
20	2.45	21.32	3.36	57.55	12.88	101.69
29	7.66	101.16	5.26	98.39	11.5	139.12
26	45.82	199.39	47.58	200.44	83.71	271.25
30	43.46	266.8	44.92	295.03	51.42	344.95
27	15.53	183.45	19.13	178.31	26.64	211.21
Average value	12.04	92.15	12.83	100.02	19.39	130.33

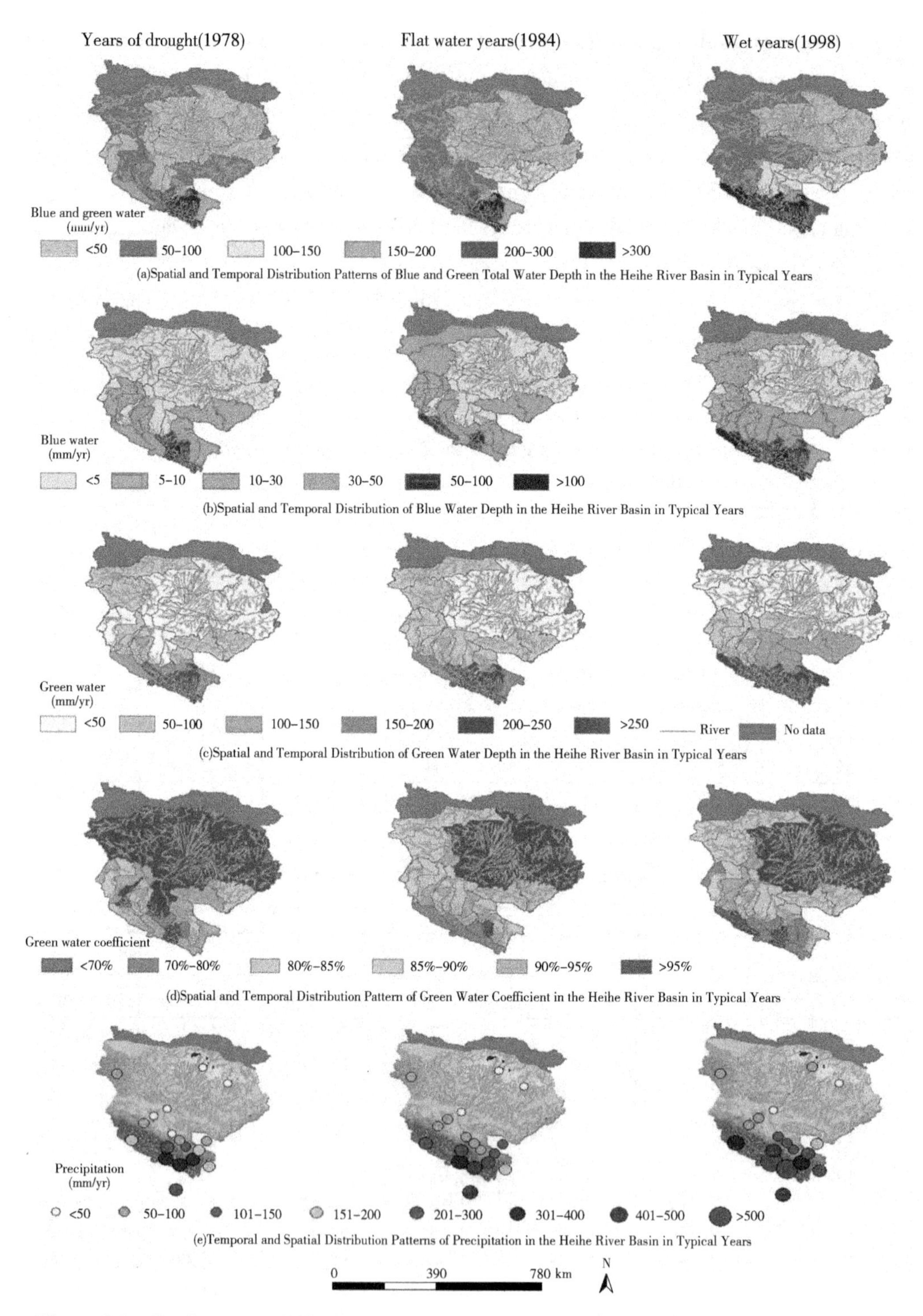

Figure 5-1　Spatio-temporal distributions of green-blue water total depth, green water and blue water depth, green water coefficient and precipitation in typical reference years in the Heihe River Basin

It can be seen from Figure 5-1(b) that the depth of blue water in the Heihe River Basin is decreasing from upstream to downstream. Blue water depth (1978, 1984, 1998 average) decreased from 98.28 mm upstream to 14.75 mm in the middle reaches and then to 6.62 mm downstream(calculated from Table 5-4 to Table 5-6). Blue water drought years(1978) to wet years(1998) significantly increased; in 1978 and 1998, blue water in the upper, middle and lower reaches are significantly different, especially in the middle and upper reaches of the region, the wet blue water significantly increased. In the upper reaches, the depth of blue water increased from 75.17 mm in 1978 to 139.25 mm in 1998 (see Table 5-4). In the middle reaches, the depth of blue water increased from 12.04 mm in 1978 to 19.39 mm in 1998(see Table 5-5). In the downstream areas, the depth of blue water increased from 1.25 mm in 1978 to 9.05 mm in 1998(see Table 5-4 to Table 5-6). The green water depth of the Heihe River Basin (1998) increased significantly in the upper and middle reaches of the dry year(1978), but decreased in the downstream areas. However, the overall depth of green water was still increasing from 1978 to 1998. The depth of the whole basin has increased from 120.57 mm in 1978 to 175.41 mm in 1998(calculated from Table 5-4 to Table 5-6). At the same time, the depth of the whole basin of green water from the overall space still show from the upstream to downstream of the decreasing trend(Figure 5-2(c)).

5.3.3 Temporal and spatial differences of green water coefficient and variation of total amount of blue water and green water in typical years

The green water coefficient in the Heihe River Basin is increasing from upstream to downstream(see Figure 5-1(d)), and the green water coefficient(the average of 1978, 1984, 1998) increased from 76.85% in the upstream to 91.66% (see Figure 5-2). At the same time, the coefficient of green water in the whole basin is obviously different in different typical years, the typical drought is 90.30% and the typical wet year is 85.41%, which indicates that the proportion of evapotranspiration in the dry year is higher than that in the wet year, The proportion of blue water to water resources is higher than that of drought years(see Figure 5-2). The greening coefficient of the wet year is lower in the middle and lower reaches of the river than in the dry years. In other words, the proportion of blue water in the Heihe River Basin is higher than that of the upper reaches of the lower rainfall and the lower reaches of the lower reaches of the rainfall. In addition, from flat water years(1984) green water coefficient, the Heihe River Basin annual utilization of water resources is 88.32% in the form of green water to participate in the hydrological cycle. It can be seen from Figure 5-2 that the total amount of blue water and green water in the Heihe River Basin is significantly larger than in the dry year(16.773 billion m^3 in 1978). This is mainly due to precipitation in the dry years significantly less than the precipitation in the wet years(see Figure 5-1(e)).

5.4 Discussion and summary of this chapter

Due to the large area of the Heihe River Basin, the middle and lower reaches of the terrain

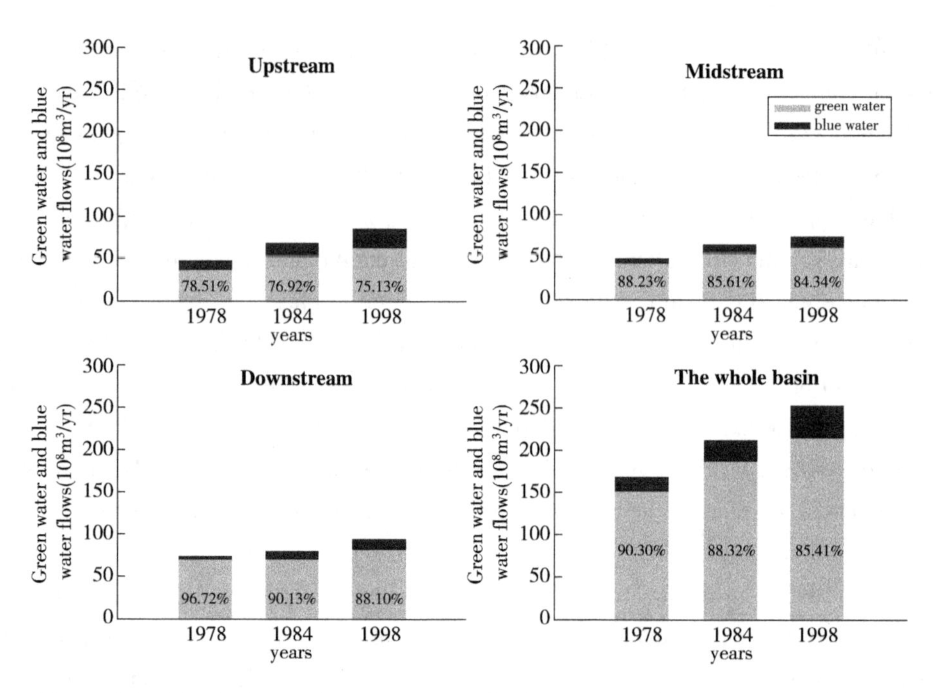

Figure 5-2 Total volume of green water and blue water and green water coefficient in typical reference years in up-mid-downstream and the entire river basin in the Heihe River Basin

and precipitation sources and regional climate affected by the atmospheric circulation (Lan Yongchao et al. ,2004;Jia wenxiong et al. ,2008;Rudi et al. ,2010) wet and wet year inconsistent. The combined depth of blue water and green water in the Heihe River Basin is the decreasing distribution from upstream to downstream, which is mainly caused by the spatial distribution of precipitation in the basin(Lan et al. ,2004;Jia Wenxiong et al. ,2008) ,but in the upper and middle reaches of some sub-watersheds, especially in the southeastern waters of the basin, which is mainly caused by the large amount of ice and snow in some sub-basins and the distribution of some lakes. (Wu et al. ,2010;Zang et al. ,2012).

The green water in the Heihe River Basin is higher in the dry years and in the lower reaches, while in the wet years and in the upper reaches of the region is relatively low. The average annual greening coefficient for the Heihe River Basin is 88. 04% (Zang et al. ,2012). In this study, the annual water year was 88. 32% ,the drought year was 90. 30% and the wet year was 85. 41%. Based on the green water coefficient of plain water, it is inferred that about 88% of the available water resources in the Heihe River Basin participate in the hydrological cycle in the form of green water. The results show that the evolution of blue-green water resources in the context of climate change and the rational management of arid and semi-arid areas inland river basin water resources, especially green water resources has important scientific significance.

In this paper, the spatial and temporal changes of blue-green water depth, the spatial-temporal changes of green water and the total amount of blue-green water in the Heihe River Basin and the upper and lower reaches of the middle and lower reaches of the Yellow River Basin are as follows: ①the total depth of the blue-green water depth decreases from 380. 46 mm upstream to 56. 32 mm downstream, and the blue water and green water depth also have the same trend, which is closely related to the upper and lower reaches of the Heihe River Basin. ②The blue-green water in the typical dry years of the Heihe River Basin was significantly lower than that in the typical wet years. The total depth of blue-green water increased from 150. 06 mm in the dry year to 231. 31 mm in the wet year, which was caused by precipitation in the Heihe River Basin. ③The green water coefficient in the Heihe River Basin is increasing from upstream to downstream, and the green water coefficient increases from 76. 85% in the upstream to 91. 66% in the downstream. At the same time, the green water coefficient (90. 30%) of the typical drought year is higher than that of the typical wet year(85. 41%). Therefore, the proportion of evapotranspiration(green water) in the dry year is obviously higher than that in the wet year, the proportion of runoff(blue water) to water resources is significantly higher than that of drought years.

Although the temporal and spatial distribution of blue-green water in different typical years has been studied, the mechanism and law of blue-green water conversion in the basin are still unclear and need further study. Strengthening research in this area will help to more rationally and effectively prevent and manage drought and floods in some areas in typical years. In particular, the study of green water and its utilization in the inland river basins in the arid regions of northwest is conducive to alleviating the water shortage crisis and the ecosystem pressure in these areas. Therefore, the spatial distribution pattern of blue-green water in the inland river basin in arid area is further deepened, and the transformation rule of blue water and green water is discussed. The key water cycle and ecological system scientific problems in arid area are discussed, and the sustainable utilization of water resources and management countermeasures, to achieve sustainable development of the basin has important theoretical and practical significance.

Chapter 6 Analysis on the Evolution Trend of Blue-Green Water in The Heihe River Basin

6.1 Research background

The impact of climate change on water resources availability has caused sustainability concerns around the world(Piao et al. ,2010;Vörösmarty et al. ,2000). With global warming and extreme weather events happening more and more frequently,water resources face serious challenges(Meehl et al. ,2000;Vörösmarty et al. ,2000). Global climate change has decreased water resources in many regions(Kundzewicz et al. ,2008),and especially in arid and ecologically fragile regions,water scarcities have become increasingly serious. Water shortages decrease both agricultural production and food security,and can also impede socioeconomic development and endanger ecosystem health(Cheng et al. ,2006;Piao et al. ,2010).

The concepts of green and blue water were proposed flrstly by Falkenmark(1995),but the concepts did not distinguish flux and volume of green water and blue water. Later on,Falkenmark and Rockström(2006) further clarify these concepts. green water is precipitation that is stored in unsaturated soil,and blue water is the water in the rivers,lakes,wetlands and shallow aquifers. From a flux point of view,green water refers to evapotranspiration,and blue water includes the liquid water flows,e. g. runoff and aquifer recharge. Green water flow dominates water uses. More than 80% of the water consumed to support global crop production comprises green water(Liu et al. ,2009),whereas grassland and forest ecosystems depend almost entirely on green water. In arid and semi-arid regions,green water plays an essential role in crop production and the provision of ecosystem services(Cheng and Zhao,2006;Falkenmark,1995). Despite the importance of green water,water resources assessments have generally emphasized blue water (the water in rivers,lakes,wetlands,and shallow aquifers),and have often ignored green water (Cheng and Zhao,2006;Falkenmark,1995). Recent studies of the two types of water have focused mainly on developing a concept or theoretical framework(Falkenmark and Rockström, 2006;Rockström et al. ,2009),developing simulation models(Faramazi et al. ,2009) and estimating quantities(Rost et al. ,2008;Schuol et al. ,2008;Liu and Yang,2009). Many studies are available for green and blue water assessment on a global scale(Rost et al. ,2008;Liu et al. ,2009),but a new focus is needed on the catchment scale(Rockström et al. ,2010). Thus far,the absence of studies on long term trends of green and blue water flows should be further strengthened.

Trend analysis is important in studies of hydrology and climatology,and has become particularly important in the context of climate change(Karpouzos et al. ,2010). Two approaches are often used for trend analysis,among others. One approach is to combine meteorological data into a hydrological model,and set several scenarios to analyze flow regimes in the context of global change. Such a scenario-analysis approach is widely used for predicting future flow regimes in the context of climate change,e. g. river flow regimes in Europe(Schneider et al. ,2013),global runoff(Piao et al. ,2007) and global water resources(Arnell,2004). The second approach is to use statistical tests to analyze past trends (changing direction, magnitude and abrupt change point) and qualitatively predict future trends(Feidas et al. ,2004;Sen,1968;Kubilius and Melichov,2008). Precise estimates of the impacts of climate change will help resource managers to respond effectively. From the perspectives of global and regional water resources management,it is also important to identify why an abrupt change occurred(Alley et al. ,2003). However,previous trend analysis has mainly focused on runoff,which is a type of blue water flow(Cheng et al. ,2007),and potential evapotranspiration(Jin and Liang,2009). Thus,there is a need for analyzing the green water flow and blue water flow regimes under global change。

In this study,we selected the Heihe River Basin as a case study(see Figure 6-1). The Heihe River,in northwestern China, is the second-largest inland river in China. The basin is in a typical arid region that is suffering from a serious shortage of water. The basin's ecosystems vary widely,ranging from mountains and associated alpine ecosystems in the south to oases in the midstream basin and deserts in the northern downstream basin(Cheng et al. ,2006). The Heihe River Basin is a strategic region because the Hexi Corridor,which passes through its middle reaches,connects the inland Xinjiang Province with the rest of northern China (Chong, 2002). In recent years, there have been many studies of the temperature, precipitation(Li et al. ,2011),and potential evapotranspiration trends in this basin(Wang,2011),but there have been no studies of the flows of both green and blue water. Therefore,a detailed and integrated analysis of the water resource trends in the basin is urgently needed to support water management for the entire river basin.

In this study,we investigated changes in four key hydrological variables(flows of blue water and green water,the total flow,and the proportion of the total accounted for by green water) since 1960 in the context of global climate change. Speciflcally,our goals were:①to analyze the above-mentioned flows on three spatial scales: the entire river basin, three regions (the upstream,midstream,and downstream basins),and the 32 individual sub-basins;and ②to quantitatively identify the dates when the flows changed abruptly and predict future trends for these flows. Based on these results,we discuss the implications for future research and for management of the basin.

6.2 Methodology

6.2.1 Data sources

The Heihe River Basin lies between 96°05′E and 104°00′E and between 37°45′N and

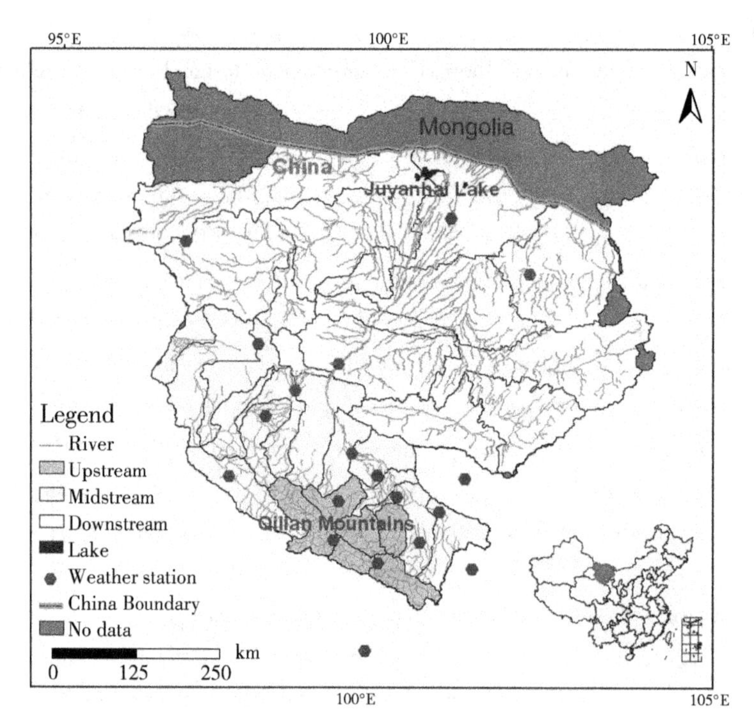

Figure 6-1 Location of the Heihe River Basin, and of the three main regions of the basin

42°40′N (see Figure 6-1). The total basin area is 0.24 × 10^6 km^2. The average elevation is above 1,200 m. The Heihe River's total length of 821 km is divided into three sections: the upstream, midstream, and downstream basins (see Figure 6-1). The upstream basin begins in the Qilian Mountains in the south and the downstream basin terminates in Juyanhai Lake (Cheng, 2002). The average precipitation for the whole basin is 185.2 mm/yr from 1960 to 2010, of which more than 70% falls from May to August (Li et al., 2010). The average annual precipitation decreases from between 200 mm/yr and 500 mm/yr in the upstream basin to between 50 mm/yr and 200 mm/yr in the midstream basin and less than 50 mm/yr in the downstream basin. Potential evaporation ranges from 1,000 mm/yr in the upstream basin to 4,000 mm/yr in the downstream basin (Li, 2009). The climate is therefore very dry, with a drought index (the ratio of potential evapotranspiration to precipitation) reaching 80 in the downstream basin. The main land cover types are desert, mountains, and oases, which together cover 98.6% of the total basin area (Cheng et al., 2006).

The flow of green water refers to actual evapotranspiration, whereas the flow of blue water is the sum of surface runoff, lateral flows, and groundwater recharge (Schuol et al., 2008). To account for the relative importance of the two flows, we deflned the green water coefflcient ($GWC = g/(b+g)$) as the ratio of the green water flow to the total flow of green water and blue water (Liu et al., 2009): Where b and g are the blue water and green water flows, respectively.

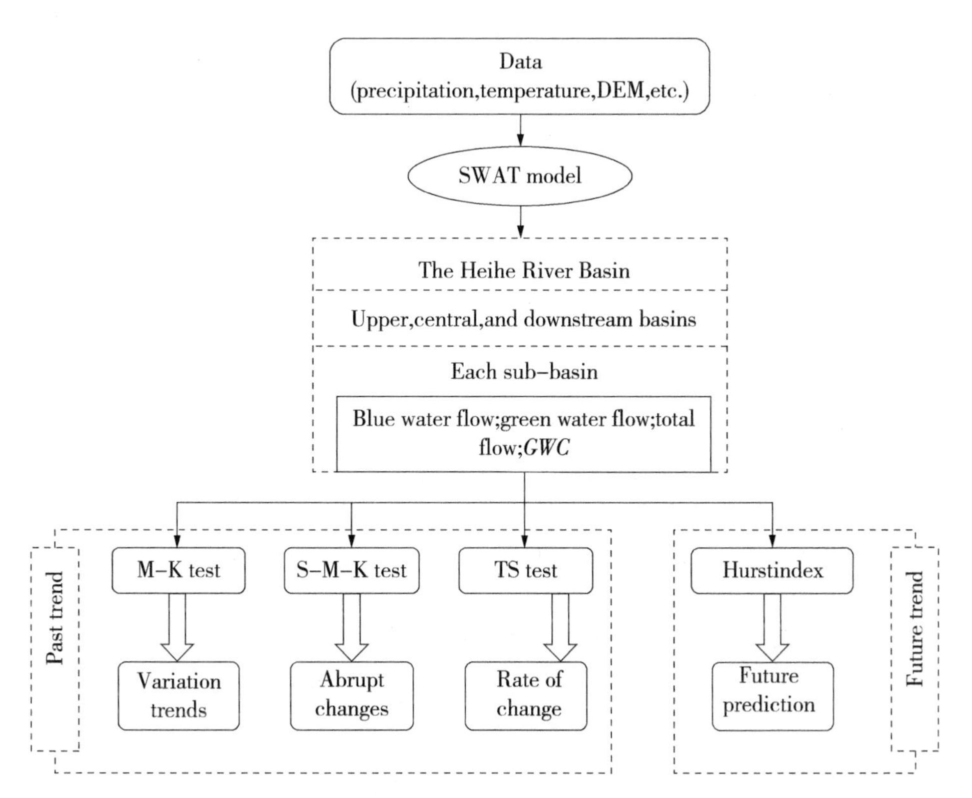

Figure 6-2 Research framework for this paper (M-K, Mann-Kendall test; SMK, sequential Mann-Kendall test; TS, Theil-Sen estimator. DEM, digital elevation model; GWC, green water coefficient (green water flow/total flow); SWAT, Soil and Water Assessment Tool)

6.2.2 Trend analysis method

The data on these flows were obtained from simulation results obtained using the Soil and Water Assessment Tool (SWAT, Arnold et al., 1998). SWAT is a semi-distributed and semi-physically based model (Neitsch et al., 2004), and it has already been successfully applied for water quantity assessments for a wide range of scales and environmental conditions in many regions all over the world (Schuol et al., 2008; Faramazi et al., 2009). This study is a followup of our previous research. We calibrated and validated the SWAT model for the simulation of the flows of green water and blue water at the whole-basin level by comparing simulated and measured discharge during 1980-2004, as shown in Zang et al. (2012). The calibration and validation performed were very satisfactory, as indicated by high values of Nash Coefflcient E_{ns} (>0.87) and R^2 (>0.90). This implies that the SWAT model is successful in modelling the blue water flows in the Heihe River Basin. There is a lack of long-term measured actual evapotranspiration (ET) for the calibration and validation of green water flows. From a long time period perspective, the sum of green water and blue water flows in a year is equal to the total precipitation. Hence, an accurate simulation of blue water flows indirectly implies a good performance of the simulation of green water flows. Nevertheless, we also checked available published literature for the measured *ET* that is only available for 2007 (Yang, 2009), and compared the measured annual *ET* with the simulated *ET* for the same year for the sub-basins that cover the monitoring sites. The comparison shows a good performance of the SWAT model, with relative errors ranging between −11%-12%. Therefore, the simulation results from the SWAT model are reliable for trend analysis. In the present study, we used the calibration parameters derived in the previous study, but expanded the simulation period to 1958-2010. The flrst two years were used as a warm-up period in the model to mitigate the effect of unknown initial conditions, and these years were then excluded from the analysis. Further information on the model simulation, calibration, and validation can be found in Zang et al. (2012). The daily precipitation and temperature data were obtained from the Heihe Data Research Group. We used data from 19 weather stations for our simulation: 7 upstream, 7 midstream, and 5 downstream stations. We used the tessellation polygon method (Bern et al., 1992) to calculate the temperature and precipitation distributions throughout the river basin. We generated tessellation polygons based on the locations of the 19 weather stations and the boundaries of the Heihe River Basin. We then used each poly-gon's area as a weight, and calculated the weighted average precipitation throughout the basin.

6.3 Analysis on the historical evolution of blue-green water

6.3.1 The changing trend of blue-green water in the Heihe River Basin

In our research, we used the Mann-Kendall (MK) trend test (Mann, 1945; Gilbert, 1987) to analyze trends of green and blue water flows in the past time series, the sequential Mann-Kendall(SMK) test (Feidas et al., 2004; Yang, 2009) to detect abrupt changes of a trend, and the Theil-Sen estimator (TS; Sen, 1968) to analyze the change magnitude of green and blue water flows, We used the Hurst index (Kubilius and Melichov, 2008) to predict future trends of green and blue water flows. Fig. 2 shows the framework for our use of these parameters. There are more than 20 approaches used for analyzing trends and abrupt changes (Karpouzos et al., 2010; Li et al., 2011). As many hydrologic or climate time series data are not normally distributed, non-parameter test were preferred over parameter test, even though they do not satisfled the number of the trend deflnition (Karpouzos et al., 2010). The MK test (Mann, 1945; Gilbert, 1987) is recommended by the World Meteorological Organization for non-parametric analysis of the signiflcance of monotonic trends of hydrological or climatological variables. It has been widely used to test for trends in hydrological and meteorological data, including precipitation, runoff, and temperature (Li et al., 2008). It has proven its ability to detect trends in hydrological time series (Burnand Hag Elnur, 2002). Several methods have been used to analyze the abrupt change of precipitation and runoff in the Heihe River basin, including the sequential MK (SMK) (Yang, 2009; Lan et al., 2004) and Pettitt method (Li et al., 2011). The advantage of the SMK test is that it cannot be influenced by a small number of outliers for sequence analysis and the calculation of such a test is simple. It has been widely used in sequence analysis for climate parameters and hydrological variables (Gilbert, 1987; Jia et al. ,2008; Yang, 2009). Therefore, the SMK is used to test assumptions about the start of a trend and detect abrupt changes in a trend (Feidas et al. , 2004; Yang,2009). The non-parametric TS method is used to estimate the magnitude of an existing trend, which isexpressed as the rate of change per year. The TS estimator can be used to estimate this rate when the trend can be demonstrated to be linear. Hurst's index (Kubilius and Melichov, 2008) (H) has a strong ability to predict future trends for a time series in relation to past trends, and it has been used to predict hydrological and climato-logical processes (Li et al. , 2008; Sakalauskiene, 2003). H ranges between 0 and 1. A value of 0. 5 means that past trends will not influence future trends. When $H > 0.5$, this means that the future trend will continue the past trend; when $H < 0.5$, this means that the future trend will differ from the past trend .

6.3.2 The changing trend of blue-green water in the middle and lower reaches of the Heihe River Basin

There are more than 20 approaches used for analyzing trends and abrupt changes (Karpouzos et al. ,2010;Li et al. ,2011). As many hydrologic or climate time series data are not normally distributed, non-parameter test were preferred over parameter test, even though they do not satisfled the number of the trend definition(Karpouzos et al. ,2010). The M-K test (Mann,1945;Gilbert,1987)is recommended by the World Meteorological Organization for non-parametric analysis of the signiflcance of monotonic trends of hydrological or climatological variables. It has been widely used to test for trends in hydrological and meteorological data, including precipitation, runoff, and temperature(Li et al. ,2008). It has proven its ability to detect trends in hydrological time series(Burn and Hag Elnur,2002). Several methods have been used to analyze the abrupt change of precipitation and runoff in the Heihe River Basin, including the sequential M-K(S-M-K) (Yang,2009;Lan et al. ,2004) and Pettitt method(Li et al. ,2011). The advantage of the S-M-K test is that it cannot be influenced by a small number of outliers for sequence analysis and the calculation of such a test is simple. It has been widely used in sequence analysis for climate parameters and hydrological variables (Gilbert, 1987; Jia et al. , 2008;Yang,2009). Therefore, the S-M-K is used to test assumptions about the start of a trend and detect abrupt changes in a trend(Feidas et al. ,2004;Yang,2009). The non-parametric TS method is used to estimate the magnitude of an existing trend, which is expressed as the rate of change per year. The TS estimator can be used to estimate this rate when the trend can be demonstrated to be linear. Hurst's index(Kubilius and Melichov,2008) (H) has a strong ability to predict future trends for a time series in relation to past trends, and it has been used to predict hydrological and climatological processes(Li et al. ,2008;Sakalauskiene,2003). H ranges between 0 and 1. A value of 0. 5 means that past trends will not in fluence future trends. When $H > 0.5$, this means that the future trend will continue the past trend; when $H < 0.5$, this means that the future trend will differ from the past trend(Sakalauskiene,2003).

The M-K test shows that the blue water and total flows generally increased signiflcantly at the entire river basin level from 1960 to 2010 ($p < 0.001$ and $p < 0.01$, respectively; Figure 6-3). As a result, *GWC* decreased signiflcantly ($p < 0.001$) at the entire river basin level. Blue water flow increased by 1.07×10^9 m^3 for the basin as a whole from 1960 to 2010, equivalent to a rate of increase of 0.21×10^9 m^3 per decade (see Figure 6-3). Total flow increased by 2.18×10^9 m^3 from 1960 to 2010, equivalent to a rate of increase of 0.44×10^9 m^3 per decade(see Figure 6-3). The green water flow for the whole basin increased by 1.11×10^9 m^3 during the study period, equivalent to a rate of increase of 0.22×10^9 m^3 per decade, but the M-K test suggested that this trend was not signiflcant at the entire river basin level(see Figure 6-3).

Based on the S-M-K test(see Figure 6-3), the blue water flow changed abruptly in 1963, whereas the green water flow changed abruptly in 1975 and 1980, the total flow changed abruptly in 1963 and 1978, and *GWC* changed abruptly in 1963. This suggests that the blue water flow and *GWC* had the same change point, but different trends. Total flow fluctuated from 1975 to 1978, but increased continuously after 1978.

In the upstream basin, the blue water, green water, and total flows generally increased signiflcantly from 1960 to 2010 (M-K test, $p < 0.001$). *GWC* generally decreased signiflcantly during this period(M-K test, $p < 0.05$). Blue water flow increased by 0.13×10^9 m^3 per decade from 1960 to 2010, versus 0.21×10^9 m^3 per decade for the green water flow. Although the above general trends are statistically signiflcant for the entire study period, different variation trends occur for blue water, green water, and total flows within the study period re flected by abrupt changes. For example, the blue water flow decreased from 1961 to 1966 and then increased from 1966 to 2010, whereas the total flow decreased from 1963 to 1966 and increased from 1966 to 2010. The green water flow started to increase from 1963, and *GWC* has decreased since 1967.

In the midstream basin, the blue water flow generally increased signiflcantly from 1960 to 2010(M-K test, $p < 0.05$). The green water and total flows also generally increased signiflcantly (M-K test, $p < 0.01$). Blue water flow in the midstream basin increased by 0.10×10^9 m^3 per decade from 1960 to 2010, versus an increase of 0.20×10^9 m^3 per decade for green water flow. However, according to Table 2, the blue water flow changed abruptly in 1964, whereas the green water flow changed abruptly in 1973 and 1981 and the total flow fluctuated from 1976 and 1979. *GWC* changed abruptly in 1964, as was the case for the blue water flow, but they had contrasting trends.

In the downstream basin, the blue water, green water, and total flows all generally decreased, but these trendswere not signiflcant. All three flows and *GWC* had similar times when the flow changed abruptly.

6.3.3 Spatial distribution of blue-green water trends in the Heihe River Basin

The trends for the blue water, green water, and total flows showed considerable spatial vari-

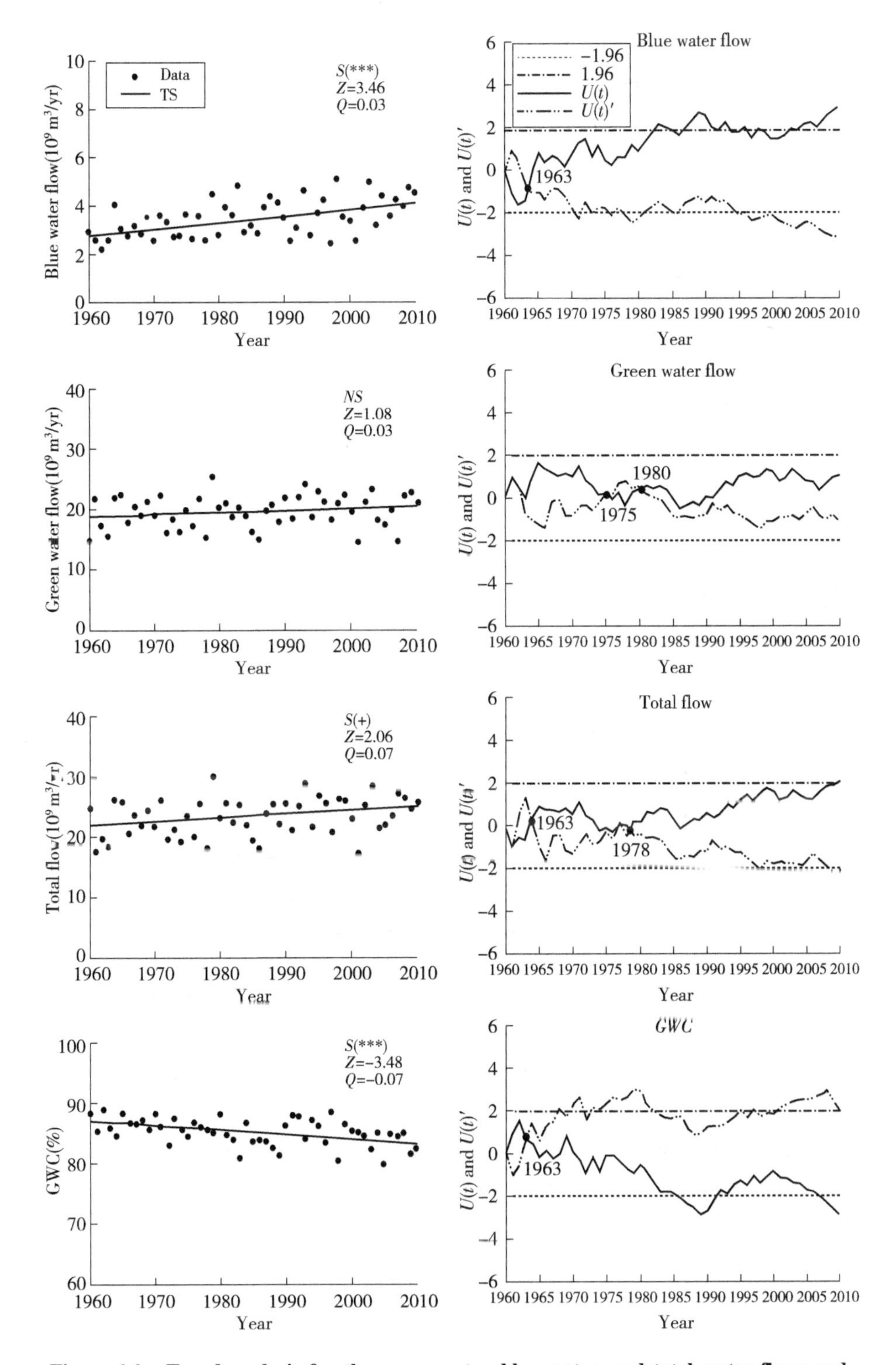

Figure 6-3　Trend analysis for the green water, blue water, and total water flows and for the green water coefficient (GWC, the proportion of the total accounted for by the green water) for the Heihe River Basin from 1960 to 2010

ation among the sub-basins(see Figure 6-4). Although the blue water flow decreased by 0.98×10^9 m^3 during the study period in the downstream basin as a whole, it increased significantly (M-K test, $p < 0.05$) in sub-basins 2, 5, 15, and 16 within the downstream basin (see Figure 6-4). The blue water flow increased significantly in the upstream and midstream basins, but decreased significantly in sub-basins 19 and 32 (M-K test, $p < 0.05$; see Figure 6-4). The green water and total flows had similar patterns at the sub-basin level, except for a significant decrease in green water flow in sub-basin 12 (M-K test, $p < 0.05$; see Figure 6-4).

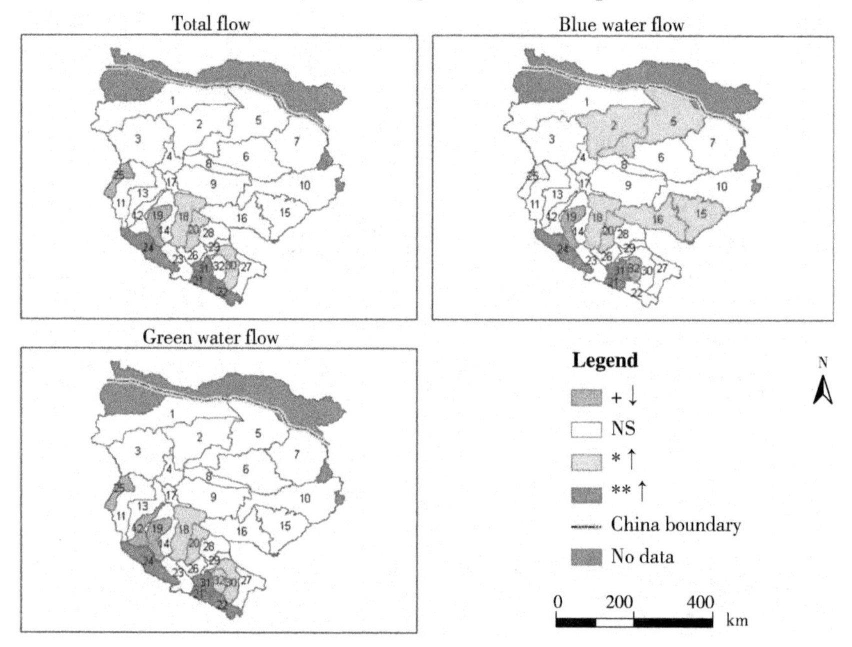

Figure 6-4 The spatial patterns for trends for the flows of blue water, green water, and their total in the Heihe River Basin from 1960 to 2010

Our results show that the trends for hydrological variables can differ among locations within a basin and among scales. Figure 6-4 shows that the total flow decreased significantly in sub-basins 19 and 25, in contrast to the general increasing trend in the midstream basin. Although the total flow had an increasing trend in the upstream and midstream basins, most sub-basins showed no significant trend, especially in the downstream basin. However, sub-basins 21, 22, 24, and 31 in the upstream and midstream basins showed significant increases in total flow (M-K test, $p < 0.01$). The total flow in these sub-basins increased by 35.4 mm, 26.8 mm, 35.7 mm and 34.0 mm, respectively, per decade from 1960 to 2010. It is therefore likely that these sub-basins strongly in fluenced hydrological processes in the upstream and midstream basins.

6.3.4 Forecast of future trends

At the whole-basin scale, the blue and total flows (but not the green water flow) and *GWC* have persistent future trends that will continue their past trends, with $H > 0.5$ (see Figure 6-5). The strength of the persistence of trends is as follows: blue water flow > *GWC* > total flow > green water flow. Based on these results, the blue water flow and total flow will keep increasing, whereas *GWC* will keep decreasing; although the green water flow may increase, the trend is not signiflcant.

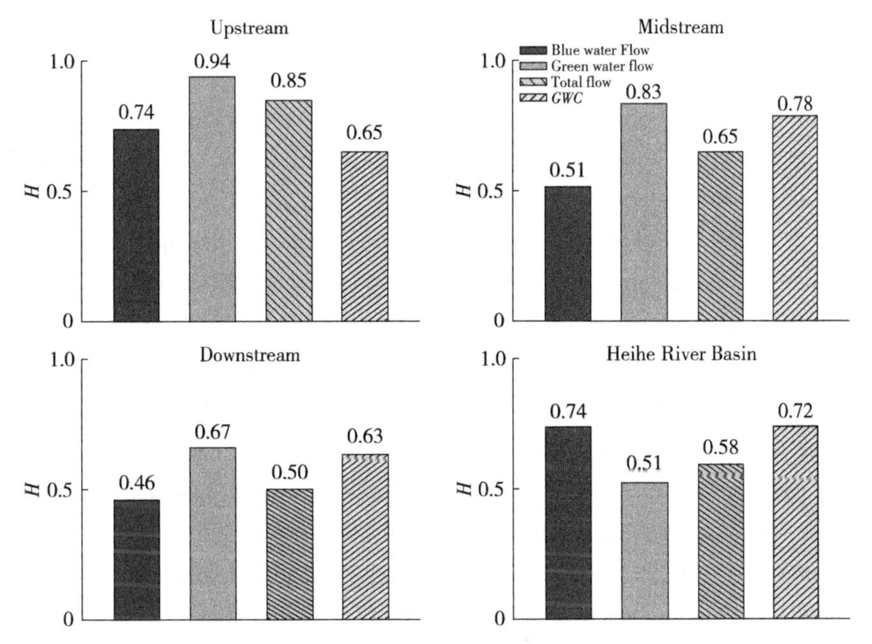

Figure 6-5 The Hurst index (*H*) for trends in the three water flows and in the green water coefficient (*GWC*, the proportion of the total flow accounted for by green water) in the Heihe River Basin. Values of $H > 0.5$ indicate that the trend from 1960 to 2010 will continue in the future. Values of $H < 0.5$ indicate that the trend will change direction

In the upstream, the hydrological variables generally have persistent future trends, with consistently $H > 0.5$ (see Figure 6-5). The strength of the persistence of trends is as follows: green water flow > total flow > blue water flow > *GWC*. Based on these results, all variables have strong persistent trends. The green water and blue water flows will keep increasing, whereas *GWC* will keep decreasing.

In midstream basins, green water flow, total flows and *GWC* have persistent future trends, with consistently $H > 0.5$ (see Figure 6-5). The green water flow and total flows will keep increasing, whereas *GWC* will keep decreasing. However, blue water flow in the midstream basin does not have an obvious persistent trend. The Hurst index is 0.51, and it is very close to the value of 0.5 that represents independent trends.

In the downstream basin, the trend for blue water flow has an anti-persistent change in the future, with $H < 0.5$; it means that the blue water flow may increase in the future in contrast to the past trend. The Hurst index for the total flow equals 0.5, indicating that the future total flow will not depend on the past trend. The green water flow and *GWC* have persistent future trends, with consistently $H>0.5$. This implies that the green water flow will increase and *GWC* will decrease in the downstream basin.

6.4 Discussion and summary

The blue water and total flows for the basin as a whole have generally increased signiflcantly from 1960 to 2010(see Figure 6-3). One main reason for this is that there has generally been signiflcantly increasing precipitation during this period, and the Hurst index values indicate that this trend is likely to continue($H=0.98$).

The trends differ for different areas of the river basin among the upstream, midstream and downstream basins. In the upstream basin, precipitation is generally high, and an increase in precipitation leads to a fast increase in runoff but a slow increase in evapotranspiration (Li, 2009). In contrast, in the downstream basin, precipitation is extremely low. An increase in precipitation will increase actual evapotranspiration but will rarely produce large amounts of runoff because of the flatter topography combined with the high potential evapotranspiration and low soil moisture content. Our analysis showed that precipitation in the upstream and midstream basins showed a signiflcant increasing trend. This would increase runoff(blue water) more than evapotranspiration(green water), leading to a lower *GWC*. In the downstream basin, precipitation did not change signiflcantly; hence, it would be logical to predict that there is no clear trend for *GWC* in that basin, which is consistent with our results. When all of these results are integrated, the clear increasing trend for the blue water flow and decreasing trend for *GWC* appear to reflect a significant increasing trend for precipitation in the upstream and midstream basins.

In the upstream basin, the blue water, green water, and total flows have generally increased signiflcantly. But these variables also had fluctuations within the study period. Both the general trends and the fluctuations are closely related to the variation of precipitation. The barrier effect of the Qilian Mountains in the upstream basin has caused different regional hydrological cycles and different precipitation sources(Jia et al., 2008). Precipitation in up-stream is influenced by different atmospheric circulations that result from the local topography, i. e. the West Paciflc Subtropical High and by the Indian Ocean Southwest Monsoon (Jia et al., 2008). The West Paciflc Subtropical High and the Indian Ocean Southwest Monsoon began to change in the 1960s, leading to an abrupt change in precipitation in 1963 (Lan et al., 2001; Jia et al., 2008). This appears to be the main reason why the blue water flow, green water and total flows fluctuated during the period of 1963-1966 in the upstream basin.

In midstream basin, the blue water flow, green water flow and total flows generally increased signiflcantly from 1960 to 2010. But they also have fluctuated in different periods, e. g. with an abrupt change in 1964 for blue water flow; with two abrupt changes in 1973 and 1980 for the green water flow. The channeling effect of the Hexi Corridor in the midstream basin has resulted in different regional hydrological cycles and precipitation sources in midstream(Lan et al., 2004). Hence, in the midstream basin, precipitation is influenced by different atmospheric circulations and topography(Lan et al. ,2004). Precipitation there is mainly influenced by the West Paciflc Subtropical High and by the Eurasian Middle-High Latitude Circulation, which have led to fluctuations of, precipitation in midstream in the 1960s and 1970s, respectively. Fluctuations influenced by the atmospheric circulations to a large extent explain the variations of the blue water flow, green water flow and total flow in the midstream basin in different periods(Lan et al. ,2004; Li et al. ,2011).

In downstream basin, the blue water flow, green water flow and total flow generally decreased from 1960 to 2010, but the variability is not signiflcant. These variables had fluctuated in the 1960s and 1970s. The precipitation in the downstream basin is mainly influenced by the Eurasian Middle-High Latitude Circulation, and the temperature there is influenced by the Mongolian High(Jia et al. ,2008; Vander Ent et al. ,2010). Precipitation did not have a signiflcant changing trend during 1960-2010, but atmospheric patterns changed abruptly from the 1960s to the 1970s(Jia et al. ,2008; Lan et al. ,2004).

Hence, precipitation in the upstream and midstream basins is most strongly affected by oceanic evaporation, whereas that in the downstream basin is most strongly affected by terrestrial evapotranspiration. The different formation mechanisms for precipitation could lead to differences among the basins in the magnitudes and trends of hydrological variables such as the three water flows. Therefore, the abrupt changes of temperature and precipitation could both have influenced the blue water and total flows, but how the different circulation systems comprehensively influence these flows and define the dates of abrupt changes in these flows will require further study. Moreover, the flows are not only influenced by atmospheric circulation, but also by local human activities, which we did not study. The impact of human activities in the Heihe River Basin will also need further research. The similar abrupt changes in the water flows, precipitation, and temperature reflect the close relationship between these three parameters in the upstream, midstream, and downstream basins.

The different precipitation patterns also caused differences in the hydrological processes at the sub-basin level. For example, the blue water flow, green water flow and total flow of sub-basin 19 had generally decreased signiflcantly at $p < 0.01$, but they had generally increased signiflcantly at $p < 0.05$ for sub-basin 18. The precipitation of the sub-basin 19 had generally decreased signiflcantly at $p < 0.01$ (Li et al. ,2011), causing the green water and blue water flows to decrease signiflcantly. However, the precipitation close to sun-basin 18 had generally increased signiflcantly at $p < 0.05$ (Li et al. ,2011), as a result, the green water and blue water flows increased.

The blue water, green water, and total water flows in the Heihe River Basin will have an increasing trend in the future. The main reason for this is that the precipitation and temperature will keep increasing throughout the Heihe River Basin as a whole.

There are several shortcomings for the current study. Firstly, the study area is a huge basin with a total area of 0.24×10^6 km^2, but only 19 weather stations provide data that can be used for statistical analysis and model simulation. Secondly, we only used one produce different conclusions. For instance, the changes of temperature detected by the S-M-K test show an abrupt change in the midstream basin in 1985, but the Pettitt method predicted a change in 1987 (Li et al., 2011; Yin, 2006). However, it is worth noting that for an examination of longterm trends (here, 50 years), a difference of only 2 years in the predicted change point is probably not very significant. Thirdly, differences between upstream, mid-stream and downstream trends are closely linked to the changing patterns of precipitation in different locations; however, other factors also play a role in influencing these differences, e. g. land use change, reservoir development, and water uses. In this study, these factors were not considered to interpret the differences. An indepth analysis is still further needed to quantitatively explain the differences among different locations by integrating models of atmospheric circulation, hydrological processes, land use change, water use and water-related infrastructure development. Last but not least, we use a hurst index to indicate future trends based upon the hydrological simulation results from the SWAT model for the past periods. There is an alternative way to analyze future trends. Recently, Intergovernmental Panel on Climate Change (IPCC) provided several simulations from General Circulation Model (GCM), depicting potential future climate under different storylines of future greenhouse gases trends. Future hydrological variables can be investigated by running the SWAT model for future projections for the Heihe River Basin by using the GCM model results from IPCC (associated with downscaling techniques) as input data and feeding these inputs to the calibrated SWAT model. Such a study is apparently out of the scope of this paper, but the comparison between the results from such a scenario analysis approach and those from the hurst index will be a very interesting exercise.

In our study, we used three statistical tests to analyze the trends for water flows and *GWC* for the entire Heihe River Basin, for three subdivisions (the upstream, midstream, and downstream basins), and for the 32 sub-basins. We also examined possible explanations for these trends.

(1) For the entire basin, the blue water and total flows have increased significantly from 1960 to 2010. These trends were most obvious in the upstream and midstream basins, especially for the blue water flow. Differences in the sources of precipitation appear to be the main cause of these trends, but the comprehensive influences of the different circulation patterns and other factors (e. g. land use change, water use) responsible for these changes will need further study. Because the blue water flow has increased more than the green water flow, *GWC* decreased at the whole-basin scale.

(2) The abrupt changes in the three water flows and in *GWC* were affected by changes in the precipitation and temperature trends. Thus, climate variability appears to be a primary cause of the abrupt changes in these variables. We also detected that the blue water flow, total flow and *GWC* had abrupt changes for the same year (1963).

(3) At the whole-basin scale, the blue water, green water, and total flows will keep increas-

ing in future, although the trend was weak for green water, whereas *GWC* will keep decreasing in the future($H = 0.51$). All hydrological factors except *GWC* will keep increasing in the upstream and midstream basins; however, only the blue water flow will continue increasing in the downstream basin. We also detected considerable spatial variation in all parameters at a sub-basin. The different topography and regional atmospheric circulation patterns changed abruptly at different times in the upstream, midstream, and downstream basins, contributing to the observed differences in the trends and dates of abrupt changes for the water flows.

(4) Our results provide insights into past and future hydrological trends throughout the Heihe River Basin and for different parts of the basin. These results provide a more comprehensive understanding of the historical and future variation of water resources in the Heihe River Basin, which will help policymakers, administrators, and researchers manage these resources in the context of global and regional climate change. Although water availability appears likely to increase in most of the basin, signiflcant decreases in some sub-basins will require particular attention from managers. In addition, the decreasing *GWC* trend suggests that in some regions, it may be necessary to pay more attention green water management for human wellbeing. Therefore, this information will also provide guidance for future studies of other inland river basins in China. Spatial variations of the changing trends imply that water managers should take speciflc adaptation measures in different regions even within a river basin to strengthen water resources management in the context of global change.

Chapter 7 Evaluation and Sustainable Analysis of Water Shortage in The Heihe River Basin

7.1 Research background

As one of the most essential natural resources, water is greatly threatened by human activities (Oki and Kanae, 2006; Postel et al., 1996; Vörösmarty et al., 2000, 2010). There are still more than 800 million people lacking a safe supply of freshwater (Ban Kimoon, 2012) and 2 billion people lacking basic water sanitation (Falconer et al., 2012). Water scarcity has been increasing in more and more countries all over the world (Yang et al., 2003). Especially in arid and semi-arid regions, nearly all river basins have serious water problems, such as rivers drying up, pollution or groundwater table decline (Joséet al., 2010; Vörösmarty et al., 2010). It is necessary to find new approaches and tools for integrated water resources management (Adeel, 2004) to help maintain a balance between human resource use and ecosystem protection (Dudgeon et al., 2006; Vörösmarty et al., 2010). New paradigms and approaches, e. g. water footprint (*WF*) and green water and blue water, have been emerging in scientific communities to promote efficient, equitable water and sustainable water uses, and these paradigms are believed to break new ground for water resources planning and management (Falkenmark, 2003; Falkenmark and Rockström, 2006; Hoekstra and Chapagain, 2007; Liu and Savenije, 2008).

WF is an indicator of water use introduced by Hoekstra (2003). It shows water consumption by source and polluted volumes by type of pollution. *WF* assessment is an human activity and water scarcity, and offer an innovative approach to integrated water resources management (Hoekstra et al., 2011). Earlier *WF* studies generally focused on five levels: process, product, sector, administrative unit, and global. At the process level, Chapagain et al. (2006) calculated the *WF* of cotton production for different processes. At the product level, Mekonnen and Hoekstra (2011) estimated the green, blue and grey *WF* of 126 crops all over the world for the period 1996-2005 with a high spatial resolution. The *WF* of pasta and pizza (Aldaya and Hoekstra, 2010) and coffee and tea (Chapagain and Hoekstra, 2007) have also been analyzed. At the sector level, Aldaya et al. (2010) calculated the *WF* of domestic, industrial and agricultural sectors in Spain and found that the inefficient allocation of water resources and mismanagement in the agricultural sector lead to water scarcity in Spain. At the national level, the *WF* of China (Liu and Savenije, 2008; Ma et al., 2006), India (Kampman et al., 2008), Indonesia (Bulsink et al., 2010), Netherlands (Van Oel et al., 2009), UK (Chapagain and Orr, 2008) and France (Ercin et al., 2012) have been assessed. At the global level, *WF* of goods and services consumed by humans have been quantified by Hoekstra and Chapagain (2007), Hoekstra and Mekonnen (2012).

Although the body of literature on *WF* has been increasing fast, there are still very few studies focusing on specific river basins (UNEP, 2011), especially for those located in arid and semi-arid regions. Assessing *WF* at a river basin level is an important step to understand how human activities influence natural water cycles, and it is a basis for integrated water resources management and sustainable water uses. *WF* assessment studies at river basin level are rare in the literature largely due to the lack of statistical data at the river basin level. Among the very few studies, input-output models have been tested to estimate *WF* at the river basin level, such as for the Haihe River Basin (Zhao et al., 2010) and for the Yellow River Basin (Feng et al., 2012). It is still necessary to test whether a bottom-up approach (Hoekstra et al., 2011) promoted by the Water Footprint Network can be successfully used for *WF* assessments for specific river basins, particularly for those in arid and semi-arid regions.

Our study aims to: ①assess *WF* at a river basin level with a bottom-up approach; and ②assess sustainability of *WF* on a monthly time step. We chose the Heihe River Basin (HRB) in inland northwest of China as a case area, and conducted a *WF* assessment by considering the agricultural (i. e. crop production and livestock production), industrial and domestic sectors. We assess the annual green *WF* and blue *WF* and compare the blue *WF* (WF_{blue}) with blue water availability (WA_{blue}) at a monthly level to pinpoint the most serious water scarce months. Located in northwest China, the Heihe River originates in the Qilian Mountains in Qinghai Province, flows through several counties in Gansu Province and Inner Mongolia, and terminates in oases in Mongolia (see Figure 7-1). The precipitation ranges from 480 mm in the upstream part of the basin to even less than 20 mm downstream. The extensive use of water in the upper and middle parts of the basin has led to a decrease in water resources downstream, causing salinization and desertification (Cheng, 2002; Feng et al., 2002; Chen et al., 2005). Previous research often pays attention to irrigation in this river basin (Chen et al., 2005; Zhao et al., 2005; Ji et al., 2006; Wang Y. et al., 2010), but a comprehensive *WF* assessment considering multiple sectors and multiple types of water (e. g. green water and blue water) has never been done before. Such an assessment is a key to better understanding the entire picture of water consumption at the river basin level, and identifying ways to improve water management.

7.2 Evaluation of water footprint in the Heihe River Basin

7.2.1 Crop blue-green water footprint

In order to assess *WF* within the HRB, we need to know the *WF* of crop production (WF_c), *WF* of livestock production (WF_l), *WF* of the industrial sector (WF_i), and *WF* of the domestic sector (WF_d). There are two types of resources: blue water (surface water and groundwater) and green water (soil water) (Liu and Savenije, 2008). Both the blue and green components of *WF* are assessed. The blue *WF* and green *WF* (WF_{blue} and WF_{green}) accounting and sus-

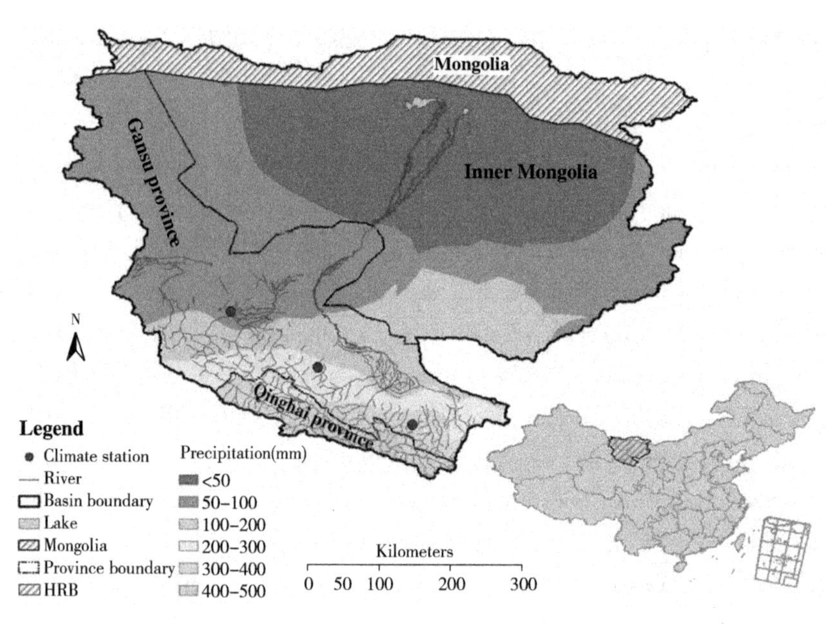

Figure 7-1 Location of the Heihe River Basin(HRB)in China

tainability assessment are mainly based on the standard methods proposed in the Water Footprint Assessment Manual(Hoekstra et al. ,2011). Because of the lack of data on pollutant discharges in the HRB, we do not include the volume of water that is used to assimilate water pollution, or grey *WF*. In this article, we only estimate *WF* within China's territory due to the lack of data in Mongolia. In addition, the area of the HRB located in Mongolia is mainly desert, while crop and livestock production and other human activities are marginal. Neglecting this area will not lead to large errors for the *WF* of the entire river basin. We assess *WF* in the HRB over 2004-2006 and use the annual and monthly results for the presentation of results.

Since many datas are not available at a river basin level, we combine statistical data for administrative boundaries(e. g. a county or a city) with spatially explicit data sets to obtain the information at the river basin level. The steps to calculate the *WF* within the HRB are depicted in Figure 7-2.

There are 15 Chinese cities or counties within or across the HRB. The statistics provide accurate information of harvested area and production of crops in these cities or counties during 2004-2006, but statistical information at river basin is not available. For these administrative regions, we need to calculate how much area is located within the HRB. With the 5 arcminute crop distribution maps from the MIRCA2000 data base from the University of Frankfurt(Portmann et al. ,2010), we can calculate the shares of crop area(both rainfed and irrigated) of one specific crop in one city or county within and outside the HRB. Combining these shares with statistical harvested area of a city or county, the crop area of all administrative regions within the HRB can be estimated. Hence, the area of each crop can be obtained at the river basin level. A similar approach is used to estimate crop production within the HRB. The results of harvested

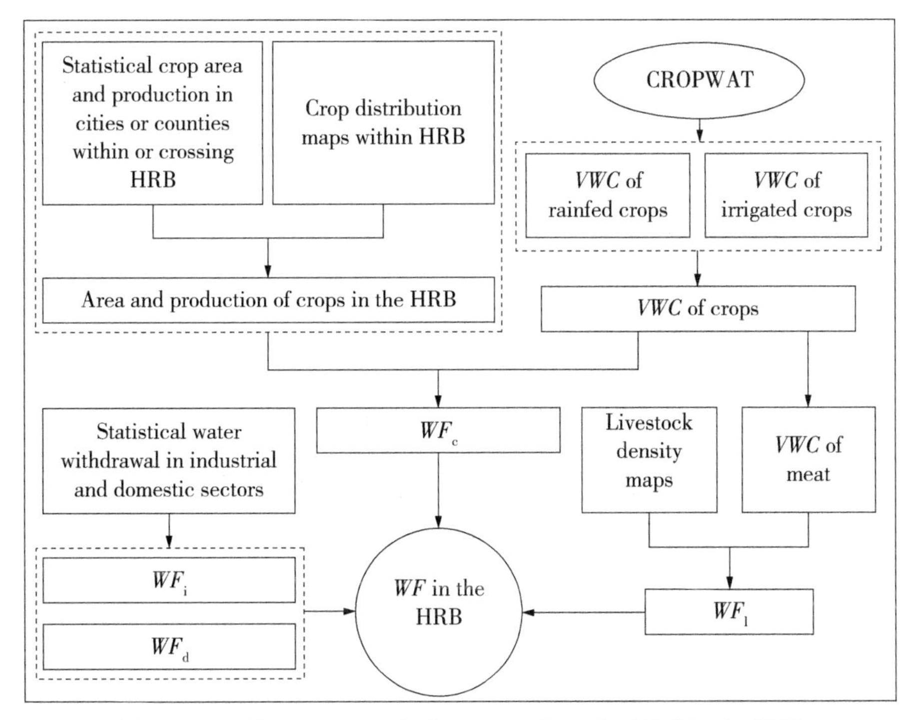

Figure 7-2 The steps to calculate water footprint(*WF*)in the HRB

area and production are shown in Table 7-1.

Table 7-1 Virtual Water Content (*VWC*), Water Footprint (*WF*) and Blue Water Proportion (*BWP*) of crop and livestock production within the HRB (2004-2006)

Crop Type	*VWC* (m^3/t)	*WF* (million m^3)	*BWP*
Wheat	826	266	64%
Maize	763	182	62%
Other cereals	1,045	368	27%
Soybean	2,216	48	72%
Starchy roots	110	10	45%
Oil crops	466	22	0
Sugar crops	94	18	0
Cotton	3,384	156	56%
Apple	855	23	34%
Other fruits	918	210	34%
Vegetables	150	111	48%
Other crops	614	225	45%
Pork	3,910	10. 32	26%
Beef	20,360	7. 62	3%
Sheep/goat	14,670	42. 87	0. 3%
Poultry	4,029	5. 01	39%

A total of 12 types of crops or crop groups were selected. Each type has its own representative crop(see Table 7-1). These duction of the animal types. Beef, sheep/goat, pork and poultry are four main animal categories in the HRB and we only consider these livestock in our calculation. The density of animals per animal category(number/km^2) is obtained from the Animal Production and Health division of FAO(2011). This dataset provides spatially explicit information on animal densities in 2005 with a spatial resolution of 3 arc-minutes. The total number of an animal in the HRB can be estimated by summing up the animal number of all grid cells within the basin.

WF_c is calculated by multiplying virtual water content(*VWC*) of each crop with its production and then summing up all crops. *VWC* is defined as the amount of water(m^3) that is needed to produce a product per unit of crop(t) during the crop growing period. The green and blue components of *VWC* are calculated as the ratio of effective rainfall(*ER*, m^3/ha) or irrigation (*I*, m^3/ha) to the crop yield(*Y*, t/ha). The *VWC* of crops is the sum of green *VWC*(VWC_{green}) and blue *VWC*(VWC_{blue}).

The CROPWAT model needs climate, crop and soil parameters to model evapotranspiration and crop irrigation requirements. Climate datas include temperature, precipitation, humidity, sunshine, radiation and wind speed. The climate datas are obtained from the New LocClim database(FAO, 2005), which provides monthly climate datas on 30-yr average(1961-1990). We selected three climate stations located in the HRB(see Figure 7-1). Crop parameters such as crop coefficients, rooting depths, lengths of each crop development stage, the planting and harvest dates are based on the studies by Allen et al. (1998) and Chapagain and Hoekstra(2004). Soil parameters include values of available soil water content, maximum infiltration rate, maximum rooting depth, and initial soil moisture depletion. Available soil water content for the HRB is retrieved from global maps from the Food and Agriculture Organization of the United Nations (FAO)(FAO, 2010b). The maximum infiltration rate depends on the soil types, which are predominantly sandy and loamy in the HRB(Qi and Cai, 2007). Because no information was available for maximum rooting depth and initial soil moisture content at the start of the growing season, default values in CROPWAT were taken(FAO, 2010a).

WF_l is calculated by multiplying *VWC* of a type of livestock meat with its production and then summing up all types of livestock types. *VWC* of meat is defined as the amount of water (m^3) that is needed to produce per unit of meat(t).

The *VWC* of meat is made up of three components: the water used to produce feed crops that the animals eat, and the drinking and processing water requirements of livestock(Mekonnen and Hoekstra, 2012). The feed of the livestock is composed of grass, rough forage and maize. In the HRB, maize needs both precipitation and irrigation, while the other crops mainly use precipitation(Zhang, 2003). The percentage of blue and green water in maize is estimated with the CROPWAT model. Drinking and processing water is dominantly "blue". We assume that feed crops are all produced within the HRB based on common practice in the HRB. The feed water requirement(FWR, m^3/kg) for an animal can be calculated by multiplying feed conversion efficiency (FCE) for a specific crop (FCE_f, kg dry mass of feed kg^{-1} of output) by the VWC of the feed crops (VWC_f, m^3/kg).

Among all crops studied, cotton has the largest VWC of 3,384 m^3/t (see Figure 7-3). Soybean also has high VWC of 2,216 m^3/t. Cereal crops in general have VWC values rang-ing from 763 to 1,045 m^3/t. The blue water proportion (BWP) is defined as the ratio of VWC_{blue} to VWC (Liu et al., 2009). Soybean has the highest BWP value of 70%, followed by wheat and maize with BWP values between 62% and 64% (Table 7-1). Sugar crops and oil crops have the lowest BWP because these crops are mainly rainfed. BWP of a crop is influenced by two factors: the share of irrigated area, and the crop characteristics, which are keys for irrigation water requirement.

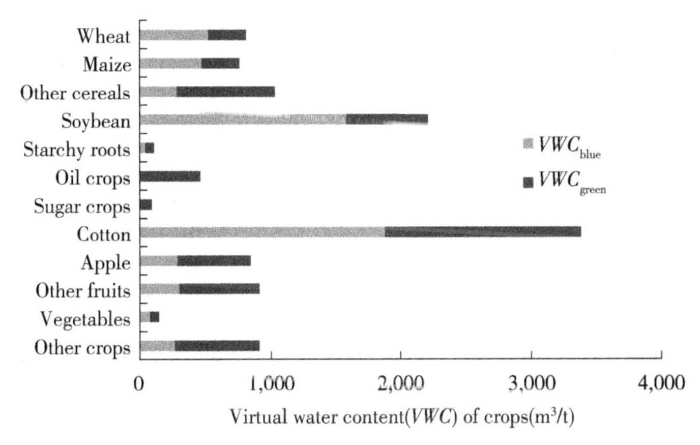

Figure 7-3 Blue and green virtual water content (VWC) of crops within the HRB

WF_c was due to the use of blue water, while the remaining 55% (896 million m^3) was from the use of green water (Figure 7-4). Cereal crops accounted for almost half of the WF_c. In particular, wheat and maize combined accounted for 27% of WF_c. Wheat and maize comprised a large share (30%) of cropland area. Cereal crops accounted for about 51% of blue WF_c and 49% of green WF_c. In particular, wheat and maize comprised 38% and 19% of blue WF_c and green WF_c, respectively. Not only in the HRB, but also for the whole China, wheat and maize are the major grain crops and account for a larger share of consumptive water use in cropland (Liu et al., 2007; Yang, 1999).

7.2.2 Livestock products blue green water footprint

Beef has the largest VWC of almost 20,000 m^3/t, followed by sheep and goat (see Table 7-2). As expected, animal meats have much higher VWC than crops. The high VWC of meat is largely due to the large feed consumption that requires a high amount of water.

Compared to crops, meat has a relatively low BWP, which ranges from less than 1% to 40% (see Table 4-2). All the four types of livestock have much higher VWC_{green} than VWC_{blue}

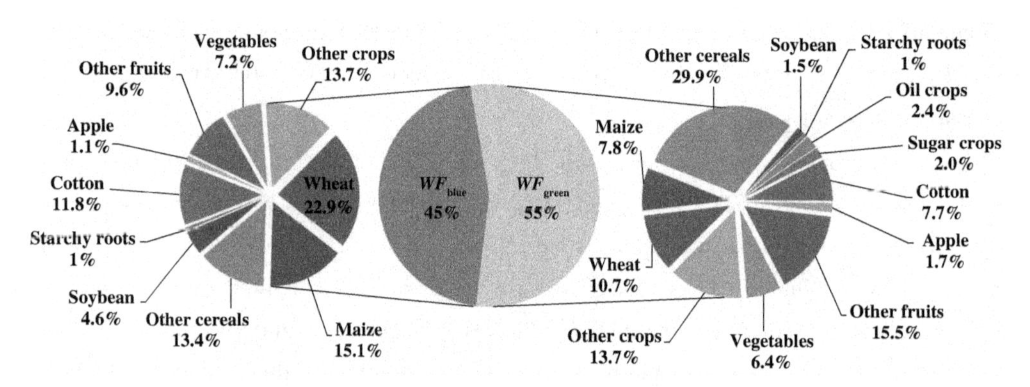

Figure 7-4 Green water and blue water footprint(WF_{green} and WF_{blue}) of crop production within the HRB over 2004-2006

compared to crops. Among the four types of meat, sheep/goat meats have the lowest BWP of 0.3%. Sheep and goat are dominantly raised in pasture land and they eat grasses in rainfed grassland without much addition to feeds such as maize. In contrast, poultry has a relative high BWP of 40%. Chicken are raised in farmers' backyards or in chicken factories, and they rely heavily on feed stuff. Hence, the BWP of chicken is significantly influenced by these feeds. The *VWC* of meats and its green and blue components are closely related to the type of feeds and animal management systems.

The average annual WF_l was 65.82 million m^3/yr in the HRB over 2004-2006. About 92% of WF_l (60.71 million m^3) was green and only 8% (5.1 million m^3) was blue (Figure 7-5). Sheep and goat accounted for over 70% of green WF_l. This is due to the large amount of meat production of sheep and goat. When checking blue WF_l, pork and poultry combined accounted for about 92%, while sheep and goat only accounted for about 4%. The low BWP of sheep and goat meats largely explains the low share of blue WF_l of sheep and goat.

7.2.3 The Heihe River Basin blue green water footprint

The average annual *WF* was 1,768 million m^3/yr in the HRB during 2004-2006 (Figure. 7-6). Almost 92% was from crop production. Livestock production accounted for 4%. The annual WF of industrial and domestic sectors in the HRB was 34 million m^3/yr and 30 million m^3/yr, respectively WF_i and WF_d combined were equivalent to WF_l. Agricultural production (crop and livestock production) was the main human activity within the HRB, and it accounted for 96% of *WF* in the HRB. For WF_c, cereal crops were the largest water user; while for WF_l, sheep and goat were the biggest water user.

In the HRB, 54% (956 million m^3/yr) of *WF* was green, while 46% (811 million m^3/yr) was blue (Figure 7-7). About 94% of WF_{green} within the HRB was related to crop production, while cereal crops contributed the largest share. WF_l only represented 6% of WF_{green}. Among WF_{blue}, crop production accounted for 91%, domestic and industrial sectors each contributed about 4%, while livestock production only accounted for less than 1%. Livestock production on-

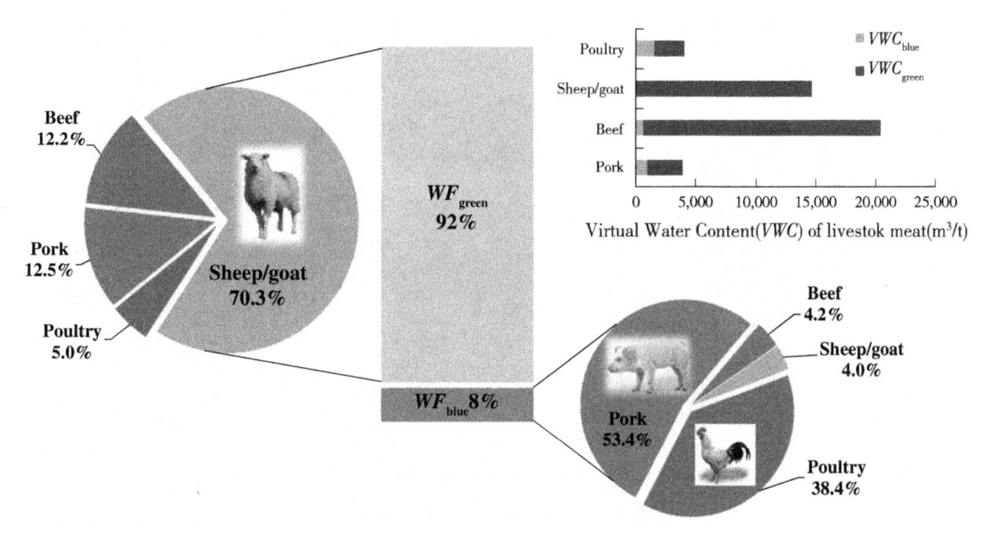

Figure 7-5 Green water and blue water footprint(WF_{green} and WF_{blue}) of livestock production within the HRB over 2004-2006

ly accounted for a marginal share of WF_{blue} because livestock in the HRB is mainly raised in pasture under rainfed conditions. Crop production, especially cereal crop production, was the main green water and blue water consumer within the HRB.

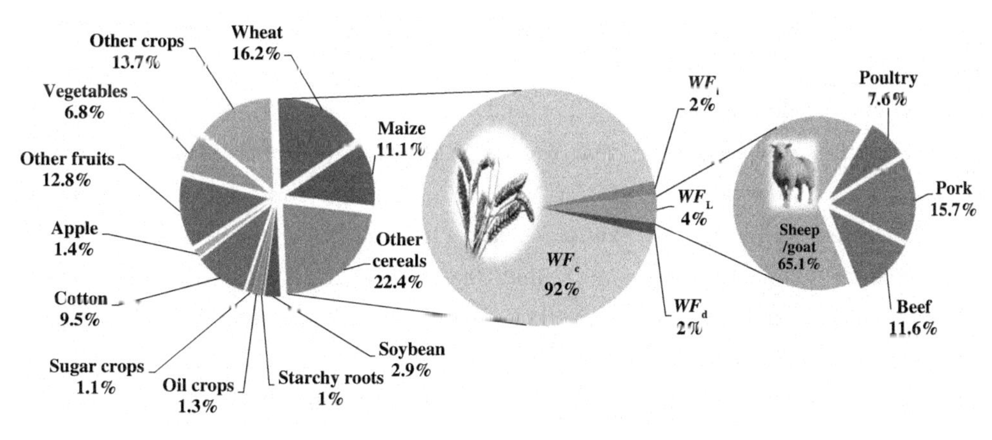

Figure 7-6 Water footprint(WF) in the HRB over 2004-2006

The per capita WF(green and blue) of the HRB is estimated to be 870 m^3/(cap · yr). According to Cai et al. (2012), in the Gansu province(the majority part of the HRB), the net virtual blue water export through food trade accounted for 10% of the total natural runoff in the basin and 25% of the total blue water use. From the water resources point of view, it is not a good solution to use precious water in arid and semi-arid regions to support a large amount of food trade. Crop pattern adjustment is a key to better water management. For different crops, the VWC of crops estimated in this paper is slightly higher than China's average values from Liu et al. (2007). One exception is cotton, and its VWC value estimated here is about twice the nation-

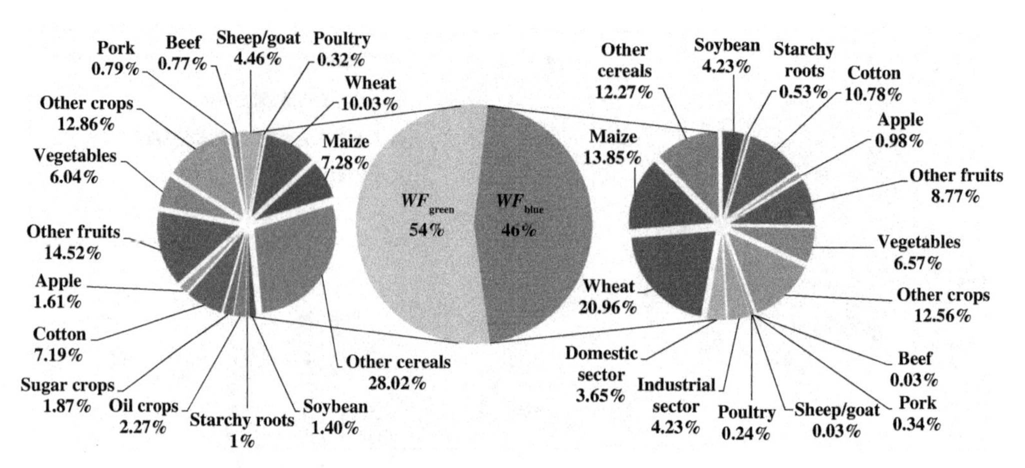

Figure 7-7 **Average annual green water and blue water footprint(WF_{green} and WF_{blue}) within the HRB over** 2004-2006

al average value. The climatic condition is one important reason for the higher *VWC* values in the HRB. The HRB is located in arid and semi-arid regions with high potential evaporating capacity. We also find that the *VWC* values of livestock products in HRB are generally higher than those reported in Chapagain and Hoekstra (2004), Liu and Savenije (2008). Especially for beef, its *VWC* value is 1.6 times the value calculated by Chapagain and Hoekstra (2004). The feed eaten by animals has higher *VWC* values in the HRB due to the dry climate conditions, leading to higher *VWC* of animal meats.

Zhang (2003) calculated *VWC* of crops and livestock in the city Zhangye located in the west of the HRB. Except for starchy roots and oil crops, the *VWC* values of all other crops and livestock reported by Zhang (2003) are very close to our results. The *VWC* of starchy roots and oil crops calculated by Zhang (2003) is much larger than ours, mainly because rainfall in the Zhangye region is lower (157-103 mm/yr) than the HRB's average level. These two types of crops mostly depend on green water rather than blue water. Low precipitation leads to high *VWC* of these two crops in the Zhangye region.

In general, the BWP of crop production in the HRB is 45%. It is much higher than the global average of 19% reported by Liu et al. (2009) and also higher than China's average of 32% (Liu et al., 2007). The HRB is an inland river basin located in arid and semi-arid northwest China. Many types of crops largely rely on irrigation during their growth period. High temperature leads to high crop water requirements, while low precipitation leads to a high dependency on irrigation in the HRB. The BWP of livestock production estimated in this study is very close to that reported in Zhang (2003).

In this study, we compare WF_{blue} with blue water availability (WA_{blue}) to indicate blue water scarcity (*BWS*) on both a yearly and monthly basis (see Figure 7-8). Natural runoff availability is high from April to September due to high precipitation in these months. WF_{blue} is also much higher from April to September than other months because crops mainly grow during these periods. The period from October to March is too cold for crops to grow. Additionally, these months have too little precipitation to support any rainfed crops.

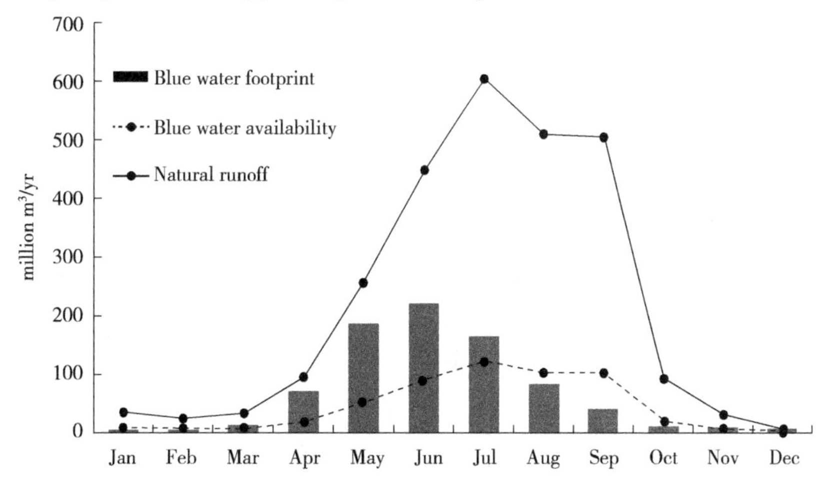

Figure 7-8 Comparison between average monthly blue water footprint and blue water availability in the HRB over 2004-2006

Hoekstra et al. (2012) provide an approach to quantify *BWS*. At a river basin level, the *BWS* is defined as the ratio of the WF_{blue} to the WA_{blue} during a certain period. It is classified into four levels: low *BWS* (< 100%), moderate *BWS* (100%-150%), significant *BWS* (150%-200%) and severe *BWS*(>200%). In the HRB, the annual WF_{blue} was 811 million m^3/yr during 2004-2006, and it was greater than the WA_{blue} of 528 million m^3/yr. The average annual *BWS* value was 154%; hence, according to the above definitions, significant *BWS* occurred on an annual basis in the HRB. WF_{blue} was 31% of the total natural runoff; hence, runoff in the HRB was significantly modified by human activities. This indicates that water consumption for human activities has exceeded the sustainable level of water availability, and human *WF* was partly met at a cost of violating environment water flows.

When comparing the monthly WF_{blue} with the monthly WA_{blue}, one can identify which months are confronted with what level of water scarcity. According to our estimate, WF_{blue} exceeded WA_{blue} in eight months of the year(see Figure 7-8). The HRB faced severe *BWS* in four months(April, May, June and December), significant *BWS* in two months(March and November), and moderate *BWS* in two months(February and July). Although high natural runoff availability occurred from April to July, WA_{blue} cannot meet human water demand, in particular for crop irrigation. From November to January, the HRB undergoes its dry season with a small amount of water available for the industrial and domestic sectors. It is clear that the environmental flow requirements are not met during two-thirds of the year. Natural runoff cannot meet human water demand and environmental flows at the same time. This leads to unsustainable water use, causing severe ecological degradation in the HRB, such as the river running dry and death of riparian vegetation(Kang et al., 2007).

7.2.4 The Heihe River Basin gray water footprint

Statistics on water use often report water withdrawal. However, we argue that *WF* is more suitable for measuring water consumption by human beings. A large part of water withdrawal will return to local water bodies and may be used again. For example, on a global scale, about 40% of agricultural water withdrawals are not consumed, but go back to downstream water bodies as return flows (Perry, 2007; Shiklomanov, 2000). Hence, water withdrawal cannot completely demonstrate human appropriation of water resources. Moreover, *WF* can quantify how much and what type (blue or green) of water is consumed by human, while the traditional statistics on water withdrawal only account for blue water. Statistics on *WF* and its "color" components (green and blue) are suggested to be reported in statistics.

Taking the HRB as an example, according to our estimate, *WF* was 1,768 million m^3/yr in 2004-2006, among which 956 million m^3/yr was green, and 811 million m^3/yr was blue. At the river basin level, there is very little statistical information on water use, even for water withdrawal. The often used water withdrawal data of 2,625 million m^3/yr in many studies (Chen et al., 2005; Zhang, 2003) are for the year of 1999. Apparently, this number includes a large amount of return flow that could further be used within the HRB. The *WF* addresses consumptive water use and its green and blue components, and shows the "real" water consumption.

Including green water in water accounting is important. Traditional water resources assessment and management mainly pay attention to blue water. In the past decades, several studies conclude that green water management should be emphasized in addition to blue water (Savenije, 2000; Liu et al., 2009). Even in arid and semi-arid regions such as the HRB, WF_{green} is still higher than the WF_{blue}, as estimated in this article. Green water plays an important role in food production. Improving green water management and green water use efficiency is key to enhancing river basin water management and to guaranteeing foodsecurity. Unfortunately, this is still an area that needs to be significantly strengthened.

7.3 Evaluation of water shortage in the Heihe River Basin

There are several shortcomings in this study. Firstly, there are no crops or livestock production datas at the river basin level. We have to calculate them based on crop or livestock distribution maps with statistics for administrative units. Such calculations can lead to errors, but this method will remain necessary when statistical data are not available at the river basin level. Our study is the first attempt for the assessment of *WF* at the HRB, and it is very difficult to validate the results obtained from the models used, such as the *VWC* of crop from the CROPWAT model. More monitoring efforts can help such validation. Secondly, for the *EFR* value, we choose 80% as a threshold based on Hoekstra et al. (2011, 2012). It is still questionable whether such a threshold can be used for river basins in arid and semi-arid regions such as the HRB. To address this issue, further efforts are still needed to study the environment flows that are required to sustain freshwater ecosystems and human livelihoods and wellbeing that depend on these ecosystems. One effective way is to set up a baseline of a "normal" water status, and evaluate the actual water requirements, especially from the local ecological systems. Third, it is very difficult to separate internal and external *WF* of the HRB and separate productive *WF* (e. g. through transpiration) and non-productive *WF* (e. g. through evaporation). Internal and external *WF* have been calculated by Cai et al. (2012) for Gansu province, which covers 43% of the HRB. The results show that the virtual water export of the agricultural products accounted for 10% of the total water resource and 25% of the total water use in the province (Cai et al., 2012). Hence, the amount of virtual water trade was quite large in such an arid region. We did not provide a comprehensive calculation of internal and external *WF* in this paper for the HRB because previous research on virtual water trade was based on input-output models, but our approach in this paper is based on the Water Footprint Network method. For the Water Footprint Network method, either the food trade data or the food consumption data should be used to estimate virtual water trade. Unfortunately, both the datasets have not yet collected successfully. As to productive/non-productive water uses, Wang and D'Odorico (2008) suggested that a focus should be on maximizing transpiration water loss and minimizing evaporation water loss. Technologies such as stable isotope analysis can be helpful to trace the water cycling processes and provide an approach for the partitioning of productive and non-productive *WF* (Wang L. et al., 2010, 2012).

There are also several factors that we did not take into account. Firstly, grey *WF* is not included due to the lack of comprehensive data on pollutant discharge. Ignoring grey *WF* will result in a conservative estimate of *WF*. Secondly, we do not calculate *WF* for the HRB outside China's boundary. However, as we have mentioned, this will not lead to large errors due to the marginal human activities for the HRB in Mongolia. Thirdly, our study did not include green water sustainability assessment. Green water plays a key role in crop and livestock production, and it is also very important to keep healthy natural ecosystems. Competition of green water between human activities and natural ecosystems will lead to different levels of green water scarcity. There are two reasons why we did not conduct a green water sustainability analysis: the lack of a standard method, and the lack of information on how much green water should be maintained for natural ecosystems. However, such analysis is an important topic and it should be further strengthened to gain indepth insights into human's intervention to green water resources. Fourthly, although we provide a first attempt to estimate *WF* for the entire the HRB, such an assessment does not take into account the spatial difference of *WF* within the river basin. Spatial heterogeneity of climate conditions and land use/cover are very sharp in the HRB with high precipitation and glaciers upstream and low precipitation and desert downstream. There is a need to compare *WF* with water availability at the sub-basin levels. This is out of the scope of this paper, but it is what will be further investigated in the next step. Fifthly, we mainly use the results of *VWC* or natural runoff from the model simulations without tracing the hydrological processes or supply chain. How detailed the calculation of *WF* should be depends on the objective of the research. To study product *WF*, it is often necessary to trace the supply chain of the product, and add up all the water needed in each chain. However, *WF* assessments at a river basin level are often based on the product *WF* results without tracing and measuring the water cycling processes. Last but not least, there is also a need to further analyze the economic and social impacts (e. g. trade, income, employment, etc.) of *WF* to enable the *WF* to become a more comprehensive indicator for decision makers.

Overall, accurate assessments of *WF* still remain a challenging task due to the complex processes of water cycles and human activities, and the lack of many important input data at a river basin level. However, it is worth extra efforts to collect more detailed information to increase the accuracy of *WF* assessment at river basin scale.

7.4 Sustainable analysis of water resources in the Heihe River Basin

The ratio of blue water footprint to blue water availability is used as a sustainable indicator of the annual or monthly shortage of blue water resources (see Figure 4-8). Due to the relatively high precipitation in April to September, the available natural runoff is higher in these months. At the same time, these months are the main growing season of the crop, so the blue water footprint is larger in this period than the other months. And October to March due to the relatively cold climate, crop growth is difficult, coupled with low precipitation is difficult to maintain the growth of rain-fed crops, so in these months the blue water footprint is low.

The annual average blue water footprint of the Heihe River Basin in 2004-2006 is 8.11×10^8 m^3/yr, which is much higher than that of blue water 5.28×10^8 m^3/yr. The average annual sustainable indicator of the basin is 154%, indicating that blue water resources are unsustainable. The blue water footprint in the Heihe River Basin is 31% of the natural runoff, and the runoff from the basin is seriously disturbed by human activities, indicating that the water consumption from human activities is far more than the sustainability of blue water use, and human water is at a great extent contrary to the law of environmental flow.

Comparing the monthly blue water footprint with the available amount of blue water, it is possible to know the sustainability of water use in each basin. According to the results of this study, the 8-month blue water footprint in the Heihe River Basin is much higher than that available for blue water (see Figure 4-8). Among them, 4 months of sustainable indicators are higher than 200% (April, May, June and December), blue water resources are in a serious unsustainable stage. Despite the high natural runoff from April to July, the amount of blue water available can not meet human needs, especially crop irrigation needs. Between November and January, the Heihe River Basin entered a dry season, with only a small portion of the water available for industrial and living sectors, indicating that two-thirds of the year was unable to meet the demand for environmental flows. Natural runoff can not simultaneously meet the water needs of humans and the environment, leading to unsustainable water use, triggering serious ecological degradation in the Heihe River Basin, such as dry rivers and river basin vegetation deaths (Kang et al., 2007).

7.5 Discussion and summary

The per capita water footprint(blue water and green water) in the Heihe River Basin is 870 m^3/(cap · yr). According to Cai Zhenhua et al. (2012), the net water exports of net blue water in Gansu Province(the main constituent areas of the Heihe River Basin) account for 10% of the total runoff of the Heihe River, and the total blue water use in the basin 25%. From a water resource perspective, it is not advisable to support such a large amount of food trade with precious water resources in arid and semi-arid areas. Adjusting the crop industry structure is the key to more effective management of water resources in the region.

The results show that the virtual water content of crops is higher than that of Liu et al. (2007). Especially cotton, the virtual water content is almost twice the average level of China. The main reason for this result is the special climatic conditions in the Heihe River Basin. The Heihe River Basin is located in arid and semiarid regions with high potential evaporation. The study also found that the virtual water content of livestock products in the Heihe River Basin was higher than that of Chapagain and Hoekstra(2004) and Liu and Savenije(2008). Especially beef products, the virtual water content is almost 1.6 times the result of Chapagain and Hoekstra(2004). The dry climatic conditions make the water content of the food in the Heihe River Basin higher, resulting in a higher level of virtual water content of meat products.

Zhang(2003) calculated the virtual water content of crop and livestock products in Zhangye City in the western Heihe River Basin. In addition to potato crops and oil crops, the virtual water content of all other crops and livestock products obtained by Zhang(2003) was similar to that of the present study. The virtual water content of potato crops and oil crops calculated by Zhang(2003) is higher than that calculated in this study, mainly because the rainfall in Zhangye is lower than that in the Heihe River Basin, and these two crops are more dependent on green water resources, low precipitation in Zhangye City led to a high level of virtual water content for both crops.

In general, the Heihe River Basin crop product has a blue water coefficient of 45%, much higher than Liu et al. (2009), the global average of 19% and the Chinese average of 32% (Liu et al., 2007). The Heihe River Basin belongs to the arid and semi-arid inland river basin in northwestern China. Most crop types depend on irrigation during the growing season. High temperatures lead to high crop water demand, while low precipitation increases the dependence of crops on irrigation in the Heihe River Basin. The blue water coefficient of livestock products calculated in this study is very similar to that of Zhang(2003).

There is still a lot of room for improvement and attention because of the lack of data and the immaturity of the method, because of the water footprint and water shortage assessment of the Heihe River Basin.

Firstly, when calculating the water footprint of the basin, due to the lack of relevant data within the basin, the results will produce some errors with the real value. However, the method used in this study is the best choice if the watershed scale statistics can not be directly obtained. For example, in the assessment of the blue-green water footprint of the basin, there is a lack of data on crop-scale and livestock products at the basin level. Based on the global distribution of crop or livestock and the statistical data of the administrative unit, the study calculates crop yields and livestock production in the Heihe River Basin, and the results will certainly differ from the actual value, but the method is already the best way to take the current situation. This study is the first attempt to evaluate the water footprint of the Heihe River Basin. It is difficult to verify the results of the model. For example, the CROPWAT model calculates the value of the virtual water content of the crop. The verification of this result can only be achieved through more monitoring and experiment.

Secondly, in the evaluation of the blue water footprint sustainability in the Heihe River Basin, 80% of the thresholds proposed by Hoekstra et al. (2011), 80% of the natural runoff, were used to maintain the ambient flow. Whether this data can be used in this arid and semi-arid area of the Heihe River Basin also needs to be verified. In order to address this problem, there is a need for more research on the relationship between the environmental flows that maintain freshwater ecosystems and the human activities that rely on these ecosystems. One of the more effective ways is to establish a baseline of "conventional" water conditions and to assess actual water demand, especially for local ecological water demand.

Thirdly, it is difficult to distinguish between the internal and external water footprints of the Heihe River Basin and the production of water footprints (such as water footprints through transpiration) and non-productive water footprints (such as water footprints by evaporation). Cai Zhenhua et al. (2012) calculated the internal water footprint and external water footprint in Gansu Province. The results show that the export of virtual water for agricultural products is about 10% of the total water resources and 25% of the total water consumption in Gansu Province (Cai Zhenhua et al., 2012). Illustrates the large amount of virtual water trade in arid regions. This study can not provide a complete calculation of the internal water footprint and external water footprint in the Heihe River Basin, since the existing internal and external water footprints are basically based on the input-output model, and this study uses the water footprint network bottom-up approach, which requires the assessment of virtual water trade based on food trade data or food consumption data, however, these two databases are not currently collected successfully. For productive/non-productive water use, Wang and D'Odorico (2008) suggested maximizing the loss of transpiration water and minimizing the loss of evaporated water. Techniques such as stable isotope analysis help to track the water cycle and provide a way to distinguish between productive and nonproductive water footprints (Wang et al., 2010a, 2012).

Fourthly, this study only considered the blue water availability, water consumption and pollution to carry out water shortage assessment. Green water is an important part of water resources, but the study did not include this part of the water shortage index system, because the assessment of green water shortage is very difficult. It is very important to incorporate the composition of green water into the water accounting. Traditional water resources assessment and management are concerned only with blue water, while ignoring the importance of green water. A great deal of research has shown that people should focus more on green water management on the basis of blue water (Savenije, 2000; Liu et al., 2009). In the arid and semi-arid areas of the Heihe River Basin, the green water footprint is much higher than the blue water footprint. Green water has a very important role in food production, increasing the management of green water and improving the effectiveness of green water use is the key to strengthen water management and ensure food security. But for now, for these aspects of research and application is still relatively small.

There are some factors that are not considered. (1) Only calculated in China within the Heihe River Basin water footprint, for the part of the Mongolian did not carry out research. Due to the lack of human activities in the Mongolian region of the Heihe River Basin, ignoring the water footprint in the Mongolian region does not have a significant impact on the overall outcome. (2) No sustainability assessment of green water. Green water is not only critical to crop production and livestock farming, but also plays an important role in maintaining the health of natural ecosystems. Human activities and natural systems for green water competition will result in varying degrees of shortage of green water resources. However, due to the lack of standard methods, the lack of maintenance of natural health needs of the green water information, so did not carry out this part of the evaluation. However, the sustainability evaluation of green water is very important because it provides a deeper understanding of human interference with green water resources. (3) Although the water footprint of the entire Heihe River Basin was evaluated for the first time, the spatial heterogeneity of the water footprint in the basin was not considered. The spatial heterogeneity of the climate and land use in the Heihe River Basin is very obvious. The upstream precipitation is high and the glaciers are more, and the downstream precipitation is low and the desert is more. It is very important to compare water footprints and available water resources in sub-watersheds, but this part is beyond the scope of this study and can be analyzed in future studies. (4) Some of the data used in this study, such as virtual water content or natural runoff, are modeled primarily by modeling, without considering hydrological processes or supply chains. The detailed degree of calculation and evaluation of water footprint is mainly based on research objectives. In order to calculate the production water footprint, it is generally necessary to trace the production supply chain and add all the water involved in the supply chain. However, in this study, the water footprint assessment at the basin scale was based on the results of the production water footprint, but did not trace and estimate the relevant water cycle processes. (5) Although the study of the Heihe River Basin as a whole, to explore the watershed water pollution situation, but can not determine the sub-basin water pollution, temporarily unable to the Heihe River Basin sub-basin analysis. (6) In the calculation of the gray water footprint in the Heihe River Basin, the gray water footprint of the Heihe River Basin should be less than that due to the lack of relevant data of the pollution source, without considering the main source of pollution in the watershed. Its actual value. Finally, there is still a need to further analyze the economic and social impacts (such as trade, income, employment, etc.) of the water footprint so that the water footprint can be a more comprehensive indicator and used by decision makers.

Water footprint evaluation method is still in the perfect stage, the water footprint model in the application process there are some urgent problems to be solved and discussed. In the future study, we should continue to think deeply about the following questions: (1) what default values should be used to account and evaluate the local water footprint in the absence of data. (2) How to select the most appropriate time. (3) The uncertainty of the selected data affects the accounting of the water footprint, and how to analyze the uncertainty and sensitivity of the water footprint under the current conditions. (4) The uncertainty of the water footprint is used to evaluate and analyze the local water footprint changes.

In general, accurate water footprint assessment and water scarcity assessment are still a challenge, as both water cycle and human activity are complex processes, and important datas at the basin scale are extremely lacking. But we still need to intensify efforts to collect more specific information to increase the accuracy of watershed scale water footprint evaluation.

Chapter 8 Conclusion and Prospect

8.1 Spatial and temporal distribution of blue-green water in the Heihe River Basin under natural conditions

The spatial and temporal distribution pattern of blue-green water in the Heihe River Basin under natural conditions shows that the blue-green water in the Heihe River Basin is gradually decreasing from upstream to downstream, which is mainly caused by the temporal and spatial differences of precipitation in the middle and lower reaches. The total amount of blue and green water in the Heihe River Basin did not show any significant change between 1980 and 2004. In the Heihe River Basin in 2000 to 2004, the average amount of blue water and green water are 220-25.5 million m^3. At the same time, the average green water ratio in the whole basin of the Heihe River Basin is over 88%, indicating that the vast majority of the water resources in the basin are in the form of green water. This is mainly due to the climate conditions, topographic features and geographical location of the Heihe River Basin. Our results show that the main factors contributing to the change of blue-green water in the Heihe River Basin from 1980 to 2004 are climate fluctuations.

This part of the study only considers the spatial and temporal distribution pattern of blue-green water under natural conditions, and does not consider the effect of human activities on the spatial and temporal distribution pattern of blue-green water in the basin. In fact, in the middle reaches of the Heihe River Basin, the impact of human activities on water resources changes is significant. In particular, the distribution of many towns in the middle reaches and the presence of large-scale irrigation areas have great influence on the distribution and change of water resources in the middle reaches of the Heihe River Basin. Thus, the current study only considers natural conditions without taking into account the effects of human activities, which are likely to cause some errors on the actual results. Therefore, considering the impact of human activities under the blue water and green water research is very necessary.

8.2 Spatial and temporal variations of blue-green water in the Heihe River Basin under the influence of human activities

The spatial and temporal distribution pattern of blue-green water in the Heihe River Basin under the influence of human activities is analyzed by climate change, land use, irrigation and comprehensive change scenarios. The results are as follows: (1) in climate fluctuation and human activities, The total amount of blue-green water in the Heihe River Basin increased significantly in the sub-basin and the southeastern sub-basin in the middle reaches of the lower reaches, while the sub-basin in the middle reaches and the sub-basin in the south-east of the lower reaches were reduced. (2) There was a significant decrease in blue-green water in the sub-basin in the western part of the middle reaches, which was mainly due to the decrease of precipitation in the area. From the whole basin, changes in climate and fluctuations in the Heihe River Basin in the past 20 years, an increase of 269 million m^3. In thc middle reaches of the urbanization process accelerated, blue water flow of the Heihe River Basin in the middle reaches of the sub-basin increased significantly. (3) As the precipitation and temperature changed from 1980 to 2005, the flow of green water in the Heihe River Basin is due to the decrease of the water flow in the middle reaches of the river basin, and the water flow in the middle reaches of the irrigation area needs to be reduced from the river and the shallow groundwater. The sub-basin in the western part of the middle reaches of the middle reaches a significant decrease, while in the lower reaches of the sub-basin there is a significant increase. There are some sub-basins in the middle and eastern part of the middle reaches of the Heihe River Basin. This is mainly due to the region in the past 20 years to accelerate the process of urbanization. The irrigation of farmland increased the source of evapotranspiration, resulting in a significant increase in the green water flow in the middle and eastern part of the middle reaches of the Heihe River Basin. (4) Climate change and volatility have led to an increase in the green water coefficient of the Heihe River Basin in the northwestern part of the lower reaches. The change of land use has led to a decreasing trend in the middle and eastern sub-basin of the middle reaches of the middle reaches, In the middle reaches of the Heihe River, there was a significant increase in the irrigation area.

Based on the analysis of the law of blue-green water change under different scenarios of climate change, land use and irrigation, this study explains the variation law of blue-green water in the Heihe River Basin under the influence of climate and human activities. However, this study still has the following shortcomings. Firstly, the relative lack of data affects further analysis of the results. Due to the lack of detailed statistical data on the national economy, the analysis of the causes of human activities needs to be further strengthened. Secondly, due to the imperfect land use data, the division of the situation is not detailed enough, the results of the article will have a certain impact.

From the different scenarios of blue and green water changes can be seen in the human activities under the influence of blue water and green water there is a certain relationship between the transformation. Blue water to green water conversion process as follows: precipitation→blue water→water and plant transpiration→green water flow; and green water flow to the blue water flow conversion process as follows: precipitation→green water→water condensation→secondary precipitation→blue water. From these two processes, it can be seen that the conversion process of blue water flow to green water flow is more direct, and the conversion of green water flow to blue water flow needs to be completed by secondary precipitation. This article simply lists the results of the conversion of blue water and green water, but the specific process of blue water and green water conversion process and the mechanism is not clear. In the real world, the processes and changes in hydrology and climate are very complex. Therefore, it is of great academic value to improve the efficiency of water resources utilization in arid and semi-arid areas by further strengthening the transformation mechanism of blue-green water and the study of hydrological processes under human conditions and climate change.

8.3 Analysis on the historical evolution of blue-green water in the Heihe River Basin

Based on the analysis of the trend of blue-green water in the whole basin, upper middle and lower reaches of the Heihe River Basin, the following conclusions are obtained: (1) between 1960 and 2010, The total amount of blue water and blue-green water increased significantly on the whole basin scale. In the upper and middle reaches of the region, the increase in blue water flow is particularly significant. The difference of precipitation sources in different regions of the watershed is the main reason for the difference of spatial trend of blue-green water. (2) Blue water flow, green water flow and the total amount of blue water and green water mutation, mainly caused by precipitation and temperature changes. The climate change in the upper and lower reaches is the main reason for the blue-green water mutation in the basin. (3) On the whole basin scale, the total amount of blue water, green water and blue and green water will continue to increase in the future, while the green water coefficient will continue to decrease. In the upper and lower reaches of the basin scale, in addition to blue water flow, other variables in the future will be consistent with the past trends. While the blue water flow in the downstream areas, the future will be an increasing trend.

In summary, we not only analyze the dynamic trend of blue water flow, green water flow, total amount of blue water and green water and green water coefficient in the whole basin scale of the Heihe River Basin, but also analyze the trend and distribution relationship between the upper and lower reaches of the basin scale and sub-basin scale. The results of the study can not only provide reference information for decision-makers, water managers and scholars to fully understand the changes of hydrology and water resources in the Heihe River Basin and its causes, but also for the scientific management of the arid and semi-arid areas of the inland river basin resources provide the necessary theoretical guidance. However, this study also has the following problems: we use only a statistical method for blue-green water, precipitation and temperature of the mutation analysis, which will accurately predict the mutation year to cause some error. So the future of the mutation year of the study, the use of the mutation year interval than the use of a single mutation year more reasonable. At the same time, the mechanism of the influence of different water vapor cycles on the trend of blue-green water change in middle and lower reaches and the mechanism of influence on mutations still need further study.

8.4 Spatial and temporal differences of blue-green water in typical years in the Heihe River Basin

By analyzing the temporal and spatial variation of blue-green water(blue-green water per unit area) ,the spatial-temporal changes of green water and the total amount of blue-green water in the Heihe River Basin and the typical middle-middle and lower reaches of the middle and lower reaches of the 1960s and 1970s,the following conclusions are drawn:(1) the blue water and green water depths also have the same trend,which is closely related to the climatic conditions of the upper,middle and lower reaches of the Heihe River. (2)The blue water and green water depths of the Heihe River Basin are gradually decreasing from upstream to downstream; The blue-green water in the typical dry years of the Heihe River Basin is significantly lower than that of the typical wet years,which is caused by the precipitation in the Heihe River Basin. (3)The green water coefficient in the Heihe River Basin is increasing from upstream to downstream. At the same time, the green water coefficient (90. 30%) in typical dry years is higher than that in typical wet years(85. 41%). Therefore,the proportion of evapotranspiration (green water) in the dry years is significantly higher than that in the wet years,and the proportion of the wet runoff(blue water) to the water resources is significantly higher than that of the dry years.

The green water in the Heihe River Basin is higher in the dry years and in the lower reaches,while in the wet years and in the upper reaches of the region is relatively low. The average annual green water coefficient of the Heihe River Basin is 88. 04% , the plain water year is 88. 32% ,the drought year is 90. 30% and the wet year is 85. 41%. Based on the green water coefficient of the plain water, it is inferred that about 88% of the available water resources in the Heihe River Basin are involved in the hydrological cycle in the form of green water. The results are of great scientific significance for the scientific understanding of the evolution of blue and green water resources in the context of climate change and the rational management of water resources,especially green water resources, in inland river basins in arid and semi arid areas. Although the temporal and spatial distribution of blue-green water in different typical years has been studied,the mechanism and law of blue-green water conversion in the basin are still unclear and need further study. Strengthening research in this area will help to more rationally and effectively prevent and manage drought and floods in some areas in typical years. In particular,the study of green water and its utilization in the inland river basins in the arid regions of northwest is conducive to alleviating the water shortage crisis and the ecosystem pressure in these areas. Therefore,the spatial distribution pattern of blue-green water in the inland river basin in arid area is further deepened,and the transformation rule of blue water and green water is discussed. The key water cycle and ecological system scientific problems in arid area are discussed,and the sustainable utilization of water resources and management countermeasures,To achieve sustainable development of the basin has important theoretical and practical significance.

8.5 Sustainability of blue-green water resources in the Heihe River Basin

This paper calculated the blue water and green water footprint and gray water footprint in the Heihe River Basin from 2004 to 2006, evaluated the water and water pollution in the river basin, analyzed the water shortage in the river basin, and the sustainability of the blue water utilization on the monthly scale. The results show that the average blue-green water footprint in the Heihe River is 17.68×10^8 m^3/yr, 54% of which is a green water footprint, 46% is a blue water footprint, 92% comes from crop products, 4% comes from livestock products, the industrial sector and the living sector each account for 2%. The average gray water footprint is 42.22×10^8 m^3/yr, of which 52% comes from 28% and 20% of the agricultural sector, industrial sector and living sector. The value of the water quality indicators (I_{blue}) and water quality indicators (I_{grey}) of Heihe River Basin are 1.3 and 1.6. respectively, which exceed the respective thresholds of 0.4 and 1.0. Therefore, the Heihe River Basin has both questions of water quantity and water quality shortage. in the 8 months of blue water footprint are much higher than the blue water available in the Heihe Riler Basin, the use of blue water resources are not sustainable.

8.6 Future prospects of blue-green water research

The study system answered how much water is the number of water in the Heihe River? How many green water? How is the distribution? How to evolve? How much water is the number of water in the Heihe River? How many green water? Water use sustainability How to solve the problem of blue and green water in the watershed, and to provide a basic data to support the Yellow River water in different scenarios, different scales and different years of blue and green water for a multi-angle interpretation, and trends. But the natural-social coupling system, the hydrological process changes are very complex and varied. The study of blue and green water in the future needs to further strengthen the study of the mechanism of the conversion of blue and green water, especially in the transformation of blue-green water under the conditions of human activities, blue-green water-virtual water cycle transformation, water and social interaction mechanism, green water And ecological service function research, and focus on exploring green water resources management approach and methods. These studies can effectively improve the efficiency of water use in arid and semi-arid areas and increase the use of water resources.

8.7 Policy recommendations

Based on the results of this book, the following four aspects of the Heihe River Basin water resources protection and management policy recommendations:

(1) To strengthen the Heihe River Basin blue water and green water basic research. The Heihe River Basin is a typical arid and semi-arid inland river basin in China. In recent years, with the implementation of the Heihe major research project of the Natural Science Foundation

of China, the hydrological research has made rapid progress. However, there are few studies on blue water and green water, and the basic research on the monitoring, simulation and management of hydrological processes in the watershed is considered. The next step should strengthen the basic research work on the blue and green water in the Heihe River Basin, blue water and green water.

(2) To strengthen the management and regulation of green water resources. Since more than 88% of the water resources in the Heihe River Basin exist in the form of green water, the emphasis on green water resources should be strengthened. Specific management and control measures can be the following: ①optimize the crop planting structure, reduce the consumption of water crops, especially in the growing season with high evapotranspiration of crops; ②reduce downstream agricultural production activities, due to the downstream high temperature, precipitation Less, potential evapotranspiration ability, the downstream areas of agricultural production will consume a lot of water resources. So reduce the downstream areas of agricultural production, can effectively reduce the consumption of green water.

(3) Management and allocation of water resources according to local conditions. The water resources in the Heihe River Basin are gradually decreasing from upstream to downstream. According to the water use in the Heihe River Basin, the middle reaches should be implemented to ensure the production of domestic water and the protection of ecological water in the lower reaches. In the whole basin water resources planning at the same time pay attention to the special circumstances of local areas. Our findings suggest that there are local sub-basin water changes and overall inconsistencies. Therefore, in the whole basin water resources management planning at the same time, we must also pay attention to the local areas of the particularity of water changes, according to local conditions for planning and management.

(4) Control the middle reaches of the oasis of agricultural scale. Oasis agriculture because of the huge water consumption, to strengthen the oasis of agricultural production scale control, can effectively save water resources, reduce the consumption of green water.

(5) To strengthen the importance of virtual water research. Due to the Heihe region climate and geographical location of the particularity, resulting in the characteristics of its resource-based water shortages. Reducing the export of water consumption products and increasing the import of water consumption products is an effective way to solve the shortage of water resources in arid and semi arid areas.

References

[1] Abbaspour K. C. ,Yang J. ,Maximov I. ,et al. Modelling hydrology and water quality in the pre-Alpine/Alpine Thur watershed using SWAT [J]. Journal of Hydrology,2007,333,413-430.

[2] Abbaspour,K. C. User Manual for SWAT-CUP,SWAT Calibration and Uncertainty Analysis Programs [R]. Swiss Federal Institute of Aquatic Science and Technology,Eawag,2007,Duebendorf,Switzerland,93pp.

[3] Alcamo,J. ,Döll,P. ,Henrichs,T. ,et al. Global estimates of water withdrawals and availability under current and future "business-as-usual" conditions [J]. Hydrolog. Sci. J. ,2003,48:339-348.

[4] Alcamo,J. ,Henrichs,T. ,Rösch,T.. World water in 2025-Global modeling and scenario analysis for the world commission on water for the 21st century [R]. Report A0002,Center for Environmental Systems Research,University of Kassel,Kurt Wolters Strasse 3,34109 Kassel,Germany,2000.

[5] Aldaya,M. M. ,Garrido,A. ,Llamas,M. R. ,et al. Water footprint and virtual water trade in Spain,Water policy in Spain[M]. CRC Press,Leiden,The Netherlands,2010,49-59.

[6] Aldaya,M. M. ,and Hoekstra,A. Y.. The water needed for Italians to eat pasta and pizza [J]. Agricultural Systems. 2010,103:351-360.

[7] Allen R. G. ,Periera L. S. ,Smith M. Crops evapotranspiration guidelines for computing crop water requirements [R]. FAO Irrigation and Drainage,1990,Paper 56.

[8] Allen,R. G. ,Pereira,L. S. ,Raes,D. ,et al. Crop Evapotranspiration:Guidelines for computing crop water requirements [M]. Irrigation and Drainage Paper,Food and Agriculture Organization of the United Nations, Rome,Italy,1998,300 pp.

[9] Alley R. B. ,Marotzke J. ,Nordhaus W. D. Abrupt climate change [J]. Science,2003,299,2005-2010(doi: 10.1126/science.1081056).

[10] Arnold J. G. ,Fohrer N. SWAT2000:Current capabilities and research opportunities in applied watershed modeling [J]. Hydrological Processes,2005,3:563-572.

[11] Arnold J. G. ,Srinivasan R. S. ,Muttiah J. R. ,et al. Large area hydrologic modeling and assessment. Part I:Model development [J]. Journal of the American Water resources Association,1998,34,73-89.

[12] Bern,M. ,Dobkin,D. ,Eppstein,D. Triangulating polygons without large angles. Proc. Annual ACM Symp [J]. Computational Geometry & apllications,1992,8:222-231.

[13] Bulsink,F. ,Hoekstra,A. Y. ,and Booij,M.. The water footprint of Indonesian provinces related to the consumption of crop products [J]. Hydrology and Earth System Sciences,2010,14:119-128.

[14] Burn D. H. ,Hag Elnur M. A. Detection of hydrologic trends and variability [J]. Journal of Hydrology, 2002,255:107-122.

[15] Chapagain,A. K. ,and Hoekstra,A. Y.. Water footprints of nations [R],Value of Water Research Report Series No. 16,UNESCO-IHE,Delft,the Netherlands,2004. available at:http://www.waterfootprint.org/Reports/Report 16Vol1.pdflast access date:29/3/2012.

[16] Chapagain,A. K. ,and Hoekstra,A. Y.. The water footprint of coffee and tea consumption in the Netherlands [J]. Ecological Economics,2007,64:109-118.

[17] Chapagain,A. K. ,Hoekstra,A. Y.. The green,blue and grey water footprint of rice from both a production and consumption perspective [R]. Value of Water Research Report Series No. 40,UNESCO ~ IHE,Delft, Netherlands,2010. www.waterfootprint.org/Reports/Report40 ~ WaterFootprintRice.pdf

[18] Chapagain, A. K., and Orr, S.. UK Water Footprint: the impact of the UK's food and fibre consumption on global water resources [R]. WWF-UK, Godalming, UK, August, 2008. available at: http://www.waterfootprint.org/Reports/Orr%20and%20Chapagain%202008%20UK%20waterfootprint-vol1.pdf. last access date: 29/3/2012

[19] Chapagain, A. K., Hoekstra, A. Y., Savenije, H. H. G., et al. The water footprint of cotton consumption: An assessment of the impact of worldwide consumption of cotton products on the water resources in the cotton producing countries[J]. Ecol. Econ., 2006, 60: 186-203.

[20] Chapagain, A. K., Hoekstra, A. Y.. The blue, green and grey water footprint of rice from production and consumption perspectives[J]. Ecol. Econ., 2011, 70: 749-758.

[21] Chen, Y., Zhang, D., Sun, Y., et al. Water demand management: a case study of the Heihe River Basin in China [J]. Physics and Chemistry of the Earth, Parts A/B/C, 2005, 30: 408-419.

[22] Eckhardt K., Haverkamp S., Fohrer N., et al. SWAT-G, a version of SWAT99.2 modified for application to low mountain range catchments [J], Physics and Chemistry of the Earth, Parts A/B/C, 2002, 27(9): 641-644.

[23] Ercin, A. E., Aldaya, M. M., Hoekstra, A. Y.. Corporate water footprint accounting and impact assessment: The case of the water footprint of a sugar-containing carbonated beverage[J]. Water Resour. Manag., 2011, 25: 721-741.

[24] Ercin, A. E., Mekonnen, M. M and Hoekstra, A. Y.. The water footprint of France [R], Value of Water Research Report Series No. 56, UNESCO-IHE, Delft, the Netherlands, 2012. available at: http://www.waterfootprint.org/Reports/Report56-WaterFootprintFrance.pdf. last access date: 29/3/2012

[25] Falconer, R. A., Norton, M. R., Fernando, H. J. S., et al. Global Water Security: Engineering the Future. National Security and Human Health Implications of Climate Change [R]. NATO Science for Peace and Security Series C: Environmental Security, Springer Netherlands, 2012: 261-269.

[26] Falkenmark M. Coping with water scarcity under rapid population growth [R]. Conference of SADC Minister, Pretoria, 1995, November 23-24.

[27] Falkenmark M. Freshwater as shared between society and ecosystems: from divided approaches to integrated challenges [J]. Philosophical Transaction, 2003, 358, 2037-2049.

[28] Falkenmark M., Rockström J. The new blue and green water paradigm: breaking new ground for water resources planning and management [J]. Journal of Water Resources Planning and Management, 2006, 3: 129-132.

[29] FAO: Global map maximum soil moisture-at 5 arc minutes, GeoNetwork grid database, Food and Agriculture Organization of the United Nations, Rome, Italy, available at: http://www.fao.org/geonetwork/srv/en/metadata.show? id = 5018&currTab = summary, 2010b. last access date: 1/3/2012

[30] FAO: New LocClim, Local Climate Estimator CD-ROM, Food and Agriculture Organization of the United Nations, Rome, Italy, available at: http://www.fao.org/nr/climpag/pub/en3_051002_en.asp, 2005. last access date: 1/3/2012

[31] Faramarzi M., Abbaspour K. C., Schulin R., et al. Modelling blue and green water resources availability in Iran [J]. Hyrological Processes, 2009, 23, 486-501.

[32] Feidas H., Makrogiannis T., Bora S. E. Trend analysis of air temperature time series in Greece and their relationship with circulation using surface and satellite data: 1955-2001[J]. Theoretical and Applied Climatology, 2004, 79: 185-208 (doi: 10.1007/s00704-004-0064-5).

[33] Feng, K., Siu, Y. L., Guan, D., et al. Assessing regional virtual water flows and water footprints in the

Yellow River Basin, China: A consumption based approach [J]. Applied Geography, 2012, 32 (2), 691-701.

[34] Fontaine R. Surface Water Quality-Assurance Plan for the Hawaii District of the U. S. GeologicalSurvey [J]. U. S. Geologeical Survey Open-File Report, Honolulu, Hawaii, 2001, 1-75 (http://hi. water. usgs. gov/publications/pubs/ofr/ofr 2001-75. pdf).

[35] Gassman P. W., Arnold J. G., Srinivasan R., et al. The world wide use of the SWAT model. Technological driver, networking impacts, and simulation trends [J]. Transactions of the ASABE, 2010, 50(4): 1211.

[36] Gassman P. W., Reyes M., Green C. H., et al. The soil and water assessment tool: historical development, applications, and future directions [J]. Transactions of the ASABE, 2007, 50, 1211-1250.

[37] Gassman P. W., Reyes M. R., Green C. H. The Soil and Water Assessment Tool: Historical development, applications, and future research directions [D], 2007, 22-37.

[38] Gerald A. M., Francis Z., Jenni E. Trends in Extreme Weather and Climate Events: Issues Related to Modeling Extremes in Projections of Future Climate Change [J]. American Meteorological Society, 2000, 81(3): 427-436.

[39] Gerten D., Hoff H., Bondeau A. Contemporary green " water flows: Simulations with a dynamic global vegetation and water balance model [J]. Physics and Chemistry of the Earth, 2005, 30, 334-338.

[40] Gilbert R. O. Statistical methods for environmental pollution monitoring [R]. Van Nostrand Reinhold, New York, 1987, 12-20.

[41] Gleick P. H. A Look at Twenty-first Century Water Resources Development [J]. Water International, 1998, 25: 127-138.

[42] Hao X. M., Chen Y. N, Xu C. C., et al. Impacts of climate change and human activities on the surface runoff in the TarimRiver basin over the last fifty years [J]. Water Resourees Management. 2008, 22(9): 1159-1171.

[43] Hargreaves G. L., Asce A. M., Hargreaves G. H., et al. Agricultural Benefits for Senegal River Basin [J]. Journal of Irrigation and Drainage Engineering, 1985, 111: 113-124.

[44] Henderson B. Exploring between site differences in water quality trends: a functional data analysis approach [J]. Environmetrics, 2006, 17: 65-80.

[45] Hoekstra, A. Y., and Chapagain, A. K.. Water footprints of nations: water use by people as a function of their consumption pattern [J]. Water Resources Management, 2007, 21: 35-48.

[46] Hoekstra, A. Y., and Mekonnen, M. M.. The water footprint of humanity [J]. Proceedings of the National Academy of Sciences, 2012a, 109: 3232-3237.

[47] Hoekstra, A. Y., Mekonnen, M. M., Chapagain, A. K., et al. Global monthly water scarcity: blue water footprints versus blue water availability [J]. PLoSONE, 2012 b, 7 (2): e32688, doi: 10. 1371/journal. pone. 0032688.

[48] Hoekstra, A. Y. (ed) Virtual water trade: Proceedings of the International Expert Meeting on Virtual Water Trade [R]. Value of Water Research Report Series No. 12, UNESCO-IHE, Delft, the Netherlands, 2003. available at: http://www. waterfootprint. org/Reports/Report12. pdf. last access date: 1/3/2012

[49] Hoekstra, A. Y., Chapagain, A. K., Aldaya, M. M., et al. The Water Footprint Assessment Manual: Setting the Global Standard [M]. Earthscan, London, UK. 2011.

[50] Hoekstra, A. Y., Chapagain, A. K.. Globalization of Water: Sharing the Planet's Freshwater Resources [M]. Blackwell Publishing, Oxford. 2008.

[51] Hoekstra, A. Y., Hung, P. Q.. Virtual water trade: a quantification of virtual water flows between nations in

relation to international crop trade [R]. Value of Water Research Report Series 11, UNESCO-IHE, Delft, the Netherlands, 2002.

[52] http://www. un. org/News/Press/docs/2012/sgsm14163. doc. htm, 2012. Last access date: 29/3/2012

[53] Jansson F. C., Rockström J., Gordon L. Linking freshwater flows and ecosystem services appropriated by people: The case of the Baltic Sea Drainage Basin [J]. Ecosystems, 1999, 351-366.

[54] Jewitt G. P. W., Garratt J. A., Calder I. R., et al. Water resources planning and modelling tools for the assessment of land use change in the Luvuvhu Catchment, South Africa [J]. Physics and Chemistry of the Earth, 2004, 15(18): 1233-1241.

[55] Jin W. X., Liang J. The temporal change of regional evapotraspiration and the impact factors in middle stream of the Heihe River Basin [J]. Journal of Arid Land Resources and Environment, 2009, 23(3): 88-92.

[56] Karpouzos D. K., Kavalieratou S., C. Trend analysis of precipitation data in Pieria Region (Greece) [J]. European Water, 2010, 30: 31-40.

[57] Kubilius K., Melichov D. On estimation of the hurst index of solutions of stochastic integral equations [R], Liet. Mat. Rink., LMD Darbai, 48/49, pp. 2008, 401-406.

[58] Kundzewicz Z. W., Mata L. J., Arnell N. W. The implications of projected climate change for freshwater resources and their management [J]. Hydrological Sciences Journal, 2008, 53(1): 3-10.

[59] Lannerstad F. Interactive comment on" Consumptive water use to feed humanity-curing a blind spot" by M. Falkenmark and M. Lannerstad [J]. Hydrol. Earth Syst. Sci. Discuss, 2005, 1: 20-28.

[60] Li N., Li J., Yu S. Effect of permafrost degradation on hydrological processes in typical basins with various permafrost coverage in western China [J]. Science China Earth Sciences, 2011, 54: 615-624.

[61] Li S. B. Satellite-based actual evapotranspiration estimation in the middle reach of the Heihe River Basin using the SEBAL method [J]. Hydrological Processes, 2010, 24: 3337-3344.

[62] Li Z. L., Shao Q. X., Xu Z. X., et al. Analysis of parameter uncertainty in semi-distributed hydrological models using bootstrap method: A case study of SWAT model applied to Yingluoxia watershed in northwest China [J]. Journal of Hydrology, 2010, 385: 76-83.

[63] Li Z. L., Xu Z. X. Detection of change points in temperature and precipitation time series in the Heihe River Basin over the past 50 years [J]. Resources Science, 2011, 33(10): 1877-1882.

[64] Li Z. L., Xu Z. X., Li J. Y. Shift trend and step changes for runoff time series in the Shiyang River basin, northwest China [J]. Hydrological Processes, 2008, 22: 4639-4646 (doi: 10. 1002/hyp. 7127).

[65] Li Z. L., Xu Z. X., Shao Q. X. i, et al. Parameter estimation and uncertainty analysis of SWAT model in upper reaches of the Heihe River Basin [J]. Hydrological Processes, 2009, 23: 2744-2753.

[66] Li Z., Zhang X. C., Zheng F. L. Assessing and regulating the impacts of climate change on water resources in the Heihe watershed on the Loess Plateau of China [J]. Science China Earth Sciences, 2010, 53: 710-720.

[67] Liu J. G., Zang C. F., Tian S. Y., et al. Water conservancy projects in China: Achievements, challenges and way forward [J]. Global Environmental Change, 2013, in press, corrected proof (Doi: org/10. 1016/j. gloenvcha. 2013. 02. 002), available online 13 March 2013.

[68] Liu J., Christian F., Yang H., et al. A global and spatially explicit assessment of climate change Impacts on crop production and consumptive water use [J]. PLoS One, 2013, 8(2): e57750, (Doi: 10. 1371/journal. pone. 0057750). Epub 2013 Feb 27.

[69] Liu J., Yang H. Consumptive water use in cropland and its partitioning: A high-resolution assessment

[J]. Science in China Series E Technological Sciences,2009b,52.

[70] Liu J. ,Yang H. Spatially explicit assessment of global consumptive water uses in cropland:Green and blue water [J]. Journal of Hydrology,2010,384:187-197.

[71] Liu J. ,Zehnder A. J. B. ,Yang,H. ,et al. Global consumptive water use for crop production:The importance of green water and virtual water [J]. Water Resources Research, 2009a, 45, (Dio: 10. 1029/2007WR006051).

[72] Liu X. ,Ren L. ,Yuan F. Quantifying the effect of land use and land cover changes on green water and blue water in northern part of China [J]. Hydrology and Earth System Sciences,2009,6(13):735-747.

[73] Liu Y. ,Zhang,B. ,Zhang Y. ,et al. Climatic change of sunshine duration and its influencing factors over-Heihe River Basin during the last 46 years [J]. Journal of Arid Land Resources and Environment,2009b, 23,72-76.

[74] Liu Z. ,Todini E. Towards a comprehensive physically based rainfall-runoff model [J]. Hydrology and Earth System Sciences,2002,6(5):859-881

[75] Liu,J. ,and Savenije,H. H. G. . Food consumption patterns and their effect on water requirement in China [J]. Hydrology and Earth System Sciences,2008,12,887-898.

[76] Liu,J. ,Zehnder,A. J. B. ,and Yang,H. . Historical trends in China's virtual water trade [J]. Water International,2007,32:78-90.

[77] Liu,J. ,YouL. ,Amini,M. ,et al. A high resolution assessment on global nitrogen flows incropland[J]. PNAS,2010,107(17):803-8040.

[78] Ma W. ,Ma Y. ,Hu Z. ,et al. Estimating surface fluxes over middle and upper streams of the Heihe River Basin with ASTER imagery [J]. Hydrology Earth System Science,2011,15,1403-1413.

[79] Ma Z. M. ,Kang S. Z. ,Zhang L. ,et al. Analysis of impacts of climate variability and human activity on stream flow for a river basin in arid region of northwest China [J]. Journal of hydrology. 2008,352(3-4): 239-249.

[80] Ma,J. ,Hoekstra,A. Y. ,Wang,H. ,et al. Virtual versus real water transfers within China [J]. Phil. Trans. R. Soc. Lond. B. 2006,361(1469):835-842.

[81] Mann H. B. Non-parametric tests against trend [J]. Econometrica,1945,13:245-259.

[82] Martin A. J. ,Williams. Human Impact on the Nile Basin:Past,Present,Future. Springer Science + Business Media B. V,2009,771-777.

[83] Mekonnen,M. M. and Hoekstra,A. Y. . A global assessment of the water footprint of farm animal products [J]. Ecosystems,2012,15:401-415.

[84] Mekonnen,M. M. ,Hoekstra,A. Y. . A global and high-resolution assessment of the green,blue and grey water footprint of wheat[J]. Hydrol. Earth Syst. Sc. ,2010,14:1259-1276.

[85] Mekonnen,M. M. ,Hoekstra,A. Y. . The green,blue and grey water footprint of crops and derived crop products[J]. Hydrology and Earth System Sciences,2011,15(5):1577-1600.

[86] Nash J. E. ,Sutcliffe J. V. . River flow forecasting through conceptual models. Part I—a discussion of principles [J]. Journal of Hydrology,1970,10,282-290.

[87] Neitsch S. L. ,Arnold J. G. ,Kiniry R. ,et al. Soil and Water Assessment Tool user's manual version 2000 [R]. Texas Water Resources Institute,2002,College Station,Texas.

[88] Neitsch S. L. ,Arnold J. G. ,Kiniry,R. ,et al. Soil and Water Assessment Tool Input/Output File Documentation Version 2005 [M]. Grassland,Soil and water research laboratory Angriculture research services & Black land research Center Texas Agricultual Experiment station,2004,50-80.

[89] Oki T. ,Kanae S. Global hydrological cycles and world water resources [J]. Science,2006,313:1068-1072.

[90] Piao S. L. ,Ciais P. e,Fang J. Y. ,et al. The impacts of climate change on water resources and agriculture in China [J]. Nature,2010,467:43-51(doi:10. 1038/nature09364).

[91] Postel S. L. ,Daily G. C. ,Ehrlich P. R. Human appropriation of renewable fresh water [J]. Science 1996, 5250:785-788.

[92] Ren L. L. ,Wang M. R. ,Li C. H. ,et al. Impacts of human activity on river runoff in the northern area of China [J]. Journal of Hydrology,2002,261(1-4):204-217.

[93] Rockström J. Future water availability for global food production:The potential of green water for increasing resilience to global change [J]. Water Resources Research,2009,45(7):W00A12(DOI:10. 1029/2007WR006767).

[94] Rockström J. ,Gordon L. Assessment of green water flows to sustain major biomes of the world:Implications for future ecohydrological landscape management [J]. Physics and Chemistry of the Earth,Part B: Hydrology,Oceans and Atmosphere,2001,11(12):843.

[95] Rockström J. ,Karlberg L. ,Wani S. P. ,et al. Managing water in rainfed agriculture—The need for a paradigm shift [J]. Agricultural Water Management,2010,4:543-550.

[96] Rockström J. ,Lannerstad M. F. M. Assessing the water challenge of a new green revolution in developing countries. Proceedings of the National Academy of Sciences of the United States of America,2007,15: 6253-6260.

[97] Rockstrtom J. On farm green water estimates as a tool for increased food production in water scarcity regions [J]. Physics and Chemistry of the Earth(B),1999,24:375-383.

[98] Rost S. ,Gerten D. ,Bondeau A. ,et al. Agricultural green and blue water consumption and its influence on the global water system [J]. Water Resources Research,2008,44(Dio:10. 1029/2007WR006331).

[99] Rudi J. ,Vander E. ,Savenije H. ,et al. Origin and fate of atmospheric moisture over continents[J]. Water Resources Research,2010,46,1-12.

[100] Sakalauskiene G. The hurst phenomenon in hydrology [J]. Environmental Research Engineering and Management,2003,3:16-20.

[101] Schuol J. ,Abbaspour K. C. ,Yang H. ,et al. Modeling blue and green water availability in Africa[J]. Water Resources Research,2008,44(Doi:10. 1029/2007WR006609).

[102] Seiler B. A,Haye L. ,Bressan. Using the standard rized precipitation index for flood risk monitoring. International. Journal of Climatology,2002,22:1365-1376.

[103] Sen P. K.. Estimates of the regression coefflcient based on Kendall's tau [J]. Journal of the American Statistical Association,1968,39:1379-1389.

[104] Shao Q. X. ,Campbell N. A. ,Modelling trends in groundwater levels by segmented regression with constraints [J]. Australian & New Zealand Journal of Statistics,2002,44:129-141.

[105] Shao Q. X. ,Li Z. ,Xu X. Trend detection in hydrological time series by segment regression with application to Shiyang River Basin [J]. Stochastic Environmental Research & Risk Assessment,2009,(DOI: 10. 1007/s00477-009-0312-4).

[106] Siriwardena L. ,Finlayson B. L. ,Me M. T. A. The impact of land use chang on catehment hydrology in large catehments:The Comet River,Central Queens land,Australia [J]. Journal of Hydrology,2006,326 (1-4):199-214.

[107] Sluter R. Interpolation methods for climate data literature review intern rapport[R]. IR 2009,04.

[108] Stonefelt M. D. ,Fontaine T. A. ,Hotchkiss R H. Impacts of climate change on water yield in the Upper Wnd River basin [J]. Journal of the American Water Resources Association,2000,36(2):321-336.

[109] Sun. G. ,Wang X. L. Estimation of surface soil moisture and roughness from multi-angular ASAR imagery in the Watershed Allied Telemetry Experimental Research(WATER)[J]. Hydrology Earth System Science,2011,15:1415-1426.

[110] Thanapakpawin P. ,Riehey J. ,Thomas D. ,et al. Effeets of land use change on the hydrologic regime of the MaeChaem river basin,NW Thailand [J]. Journal of Hydrology,2006,3 34:215-230.

[111] Timo S. ,Määttä A. ,Anttila P. Detecting trends of annual values of atmospheric pollutants by the Mann-Kendall test and Sen' s slope estimates—the excel template application MAKESENS. Finnish Meteorological Institute [J]. Air Quality Research,2002,31(1456-789X):1-26.

[112] USGS/EROS,A. G. V. T. M. E. N. L. :ASTER Global DEM Validation Summary Report [R],2009,11-20.

[113] Vörösmarty C. J. ,Green P. ,Salisbury J. Global water resources:vulnerability from climate change and population growth [J]. Science,2000,289(5477):284-288.

[114] Wang G. X. ,Cheng G. D. ,Du M. Y. . The impacts of human activity on hydrological processes in the arid zones of the Hexi Corridor,northwest China,in the past 50 years [R]. Water Resources Systems—Water vailability and Global Change(Proceedings of symposium HS02a held during IUGG2003 al Sapporo. July 2003). IAHS Publ. no. 280,2003,93-103.

[115] Wang X. P. Analysis of temporal trends in potential evaportranspiration over Heihe River Basin. Water Resource and Environmental Protection(ISWREP)[R]. International Symposium on,20-22 May 2011, 15-20.

[116] Wouter B. ,Rolandó C. ,Dert D. B. ,et al. Human impact on the hydrology of the Andean pa'ramos [J]. Earth Seienee Reviews,2006,79:53-72.

[117] Wu J. k. ,Ding Y. ,Yang X. ,et al. Spatio-temporal variation of stable isotopes in precipitation in the Heihe River Basin,northwestern China [J]. Environ Earth Science,2010,61:1123-1134.

[118] Xu Y. ,Ding Y. H. ,Zhao Z. C. A scenario of seasonal climate change of the 21st Century in northwest China [J]. Climatic and Environmental Research,2003,8(1):19-26.

[119] Yang J. ,Reichert P. ,Abbaspour K. C. ,et al. Comparing uncertainty analysis techniques for a SWAT application to Chaohe Basin in China [J]. Journal of Hydrology,2008,358,1-23.

[120] Yang J. ,Reichert P. ,Abbaspour K. C. ,et al. Hydrological modelling of the Chaohe Basin in China:Statistical model formulation and Bayesian inference [J]. Journal of Hydrology,2007,340,167-182.

[121] Yin Y. Y. Vulnerability and Adaptation to Climate Variability and Change in Western China [R]. A Final Report Submitted to Assessments of Impacts and Adaptations to Climate Change(AIACC), Project No. AS 25 Published by the International START Secretariat 2000 Florida Avenue,NW Washington,DC 20009 USA(www. start. org),2006,22-28.

[122] Zang C. F. ,Liu J. , Van der Velde M. ,et al. Assessment of spatial and temporal patterns of green and blue water flows under natural conditions in inland river basins in Northwest China [J]. Hydrology Earth System Science,2012,16(8):2859-2870.

[123] Zang,C. F. ,Liu,J. ,M. van der Velde,et al. Assessment of spatial and temporal patterns of green and blue water flows in inland river basin in northwest China [J]. Hydrology and Earth System Sciences Discuss,2012,9:3311-3338,doi:10. 5194/hessd-9-3311-2012.

[124] Zarate,E. ,et al. WFN grey water footprint working group final report:A joint study developed by WFN

partners [R]. Water Footprint Network, Enschede, Netherlands, 2010.

[125] Zhang Y. L. , Xia J. Assessment of dam impacts on river flow regimes and water quality: a case study of the Huai River Basin in P. R. China [J]. Journal of Chongqing University (English Edition), 2008, 04: 261-276.

[126] Zhang, D. 2003. Virtual water trade in China with a case study for the Heihe River Basin. Master thesis, UNESCO-IHE, Delft, the Netherlands.

[127] Zhao, C. , Nan, Z. , and Cheng, G. . Methods for estimating irrigation needs of spring wheat in the middle Heihe basin, China [J]. Agricultural Water Management, 2005, 75: 54-70.

[128] Zhao, X. , Yang, H. , Yang, Z. , et al. Applying the input-output method to account for water footprint and virtual water trade in the Haihe River Basin in China [J]. Environmental Science & Technology, 2010, 44: 9150-9156.

[129] Zhou J. , Hu B. , Cheng G. D. , et al. Development of a three-dimensional watershed modelling system for water cycle in the middle part of the Heihe Rivershed, in the west of China [J]. Hydrological Processes, 2011, 25: 1964-1978.